A.D.A.M. Interactive Anatomy 4

Student Lab Guide

THIRD EDITION

Mark Lafferty

Samuel Panella

DELAWARE TECHNICAL AND COMMUNITY COLLEGE

PEARSON

Benjamin
Cummings

San Francisco   Boston   New York
Cape Town   Hong Kong   London   Madrid   Mexico City
Montreal   Munich   Paris   Singapore   Sydney   Tokyo   Toronto

Publisher: Daryl Fox

Development Manager: Claire Alexander

Acquisitions Editor: Serina Beauparlant

Assistant Editor: Christina Pierson

Managing Editor: Wendy Earl

Production Supervisor: David Novak

Copyeditor: Michelle Gossage

Cover design: Yvo Riezebos

Composition: The Left Coast Group, Inc.

Manufacturing Supervisor: Stacey Weinberger

Executive Marketing Manager: Lauren Harp

Library of Congress Cataloging-in-Publication Data

Lafferty, Mark.

   A.D.A.M. Interactive Anatomy student lab guide / Mark Lafferty, Samuel Panella.—3rd ed.

     p. cm.

   ISBN 0-8053-5911-7 (pbk.)

1. Human anatomy—Computer simulation—Problems, exercises, etc. 2. Human anatomy—
Computer simulation—Study guides. I. Title: ADAM Interactive Anatomy student lab guide.
II. Panella, Samuel. III. Title.

   QM32.L34 2005

   611'.0076—dc22

                                                          2004017722

PEARSON

Benjamin
Cummings

ISBN 0-8053-5911-7

4 5 6 7 8 9 10–MAL–08 07 06

www.aw-bc.com

Preface

When the first edition of this lab guide was published, we were very excited about obtaining feedback from all users of our manual, but most particularly, from our own students. During these past years, they have provided us with numerous suggestions, and we have attempted to incorporate them into this edition. We have also improved the guide by responding to specific suggestions that have been made by a special group of reviewers. We hope that our efforts make this third edition more accurate and user-friendly.

Student Lab Guide Features

This lab guide provides the following features to help you get the most out of AIA.

Introduction

The Introduction includes a detailed guide designed to facilitate your work with AIA. This section will familiarize you with the program, the different ways you can view the human anatomy in the program, and the tools available to manipulate the images. In addition, a brief drill has been added at the end of the Introduction to test your understanding of how to use the various components of AIA.

Objectives

Each chapter begins with a series of objectives to give you both a broad overview of the chapter and specific content information to be mastered.

Exercises

Each chapter contains three kinds of questions. Identification questions lead you to label diagrams that correspond to images in AIA. Short-answer questions require you to find the answer somewhere within the program, and Beyond AIA questions are designed to test your understanding of a concept and increase your understanding of the interrelationships between systems.

Beyond AIA

Throughout the manual, the Beyond AIA icon **Beyond A I A** will appear in the margin to identify questions that go further than the scope of the software and require an outside source for an answer. These questions may require you to refer to your textbook, a dictionary, or class notes.

Acknowledgments

We would like to sincerely thank the following reviewers for their suggestions in helping to develop this workbook: Dwane Aulner, Community College of Southern Nevada; Stephen Burnett, Clayton College & State University; Darrell Davies, Kalamazoo Valley

Community College; Susan Gilmore, Cayuga Community College; Patricia Itaya, University of Mobile; Robert Jones, Adelphi University; Marilyn Kehoskie, Nash Community College; Howard Spector, Oakland University; and Judy Wallace, Middlesex Community College.

Our continued thanks goes to the people at Benjamin Cummings: Daryl Fox, Publisher; Claire Alexander, Development Manager; Christina Pierson, Assistant Editor; and David Novak, Production Editor, all of whose advice, expertise, and good nature kept us on track while we were developing this manuscript.

We welcome your comments and suggestions about this lab guide.

MARK A. LAFFERTY, PH.D., M.ED.   SAMUEL J. PANELLA, M.S., B.S.
Exercise Science Program Coordinator   Science Department Chairperson
Delaware Technical and Community College
333 Shipley Street
Wilmington, DE 19801
e-mail: lafferty@college.dtcc.edu and/or spanella@college.dtcc.edu

A.D.A.M.® Software Products
A.D.A.M.® Interactive Anatomy

A.D.A.M.® Interactive Anatomy is a multimedia software tool that dramatically enhances the study of human anatomy and anatomy-related topics at all levels of education. This flagship product provides powerful tools and search capabilities, offering the end user unparalleled access to over 20,000 anatomical structures in four different views. 3D images based on the Visible Human data set, cadaver photographs from the Bassett collection, pinned images, and Slide Show, a built-in curriculum integration and authoring tool, augment A.D.A.M.® Interactive Anatomy's core of award-winning digital medical illustrations. One-button Internet access provides solutions for distance learning as well as offering seamless integration of the Internet and its capabilities.

A.D.A.M.® Benjamin Cummings Interactive Physiology® (IPweb)

Interactive Physiology® offers an exciting interactive exploration of the physiological concepts and processes that are difficult to teach and learn. This comprehensive program is designed to clarify the complex physiological processes that are most difficult to understand. Codeveloped by Benjamin Cummings and A.D.A.M.® Software, Inc., this product uses richly detailed graphics, animation, sound, and interactive quizzes to bridge the gap between simply memorizing a concept and truly understanding it.

Available Topics: Muscular System, Nervous System I and II, Cardiovascular System, Respiratory System, Urinary System, and Fluid, Electrolyte and Acid/Base Balance.

All seven topics are now available on one CD-ROM; contact your Benjamin Cummings sales representative to order. You may also purchase the online version of Interactive Physiology®, IPweb, at www.interactivephysiology.com.

Contents

Chapter 4 The Special Senses 87

Introduction

Welcome to the A.D.A.M.® *Interactive Anatomy Student Lab Guide*. This lab guide is designed to be used with the software program A.D.A.M.® Interactive Anatomy (AIA) and is appropriate for students in the health and life sciences curricula.

The AIA multimedia software program can help you learn about human anatomy—A.D.A.M.® stands for Animated Dissection of Anatomy for Medicine. This dynamic program contains hundreds of layers of precise anatomical illustrations that are combined with an interactive toolbox to let you examine and identify hundreds of structures in the human body. Because of its flexibility, AIA provides many ways for you to study. You can:

- dissect the human body, layer by layer, in six different views and by gender
- study pinned anatomical images
- explore the brain, ear, heart, skull, lungs, eye, and female and male reproductive systems in 3D views
- explore the 12 major systems of the body
- point and click to identify structures

Learning to Use AIA

This section contains six exercises to demonstrate useful features and tools for studying anatomy using AIA:

- Introduction Exercise 1: Opening the Program
- Introduction Exercise 2: Dissectible Anatomy
- Introduction Exercise 3: Atlas Anatomy
- Introduction Exercise 4: 3D Anatomy
- Introduction Exercise 5: Clinical Animations
- Introduction Exercise 6: Using the Find Feature
- Introduction Exercise 7: Dissectible Anatomy Drill

Introduction Exercise 1   Opening the Program

➤ *Open AIA by double-clicking the Interactive Anatomy icon.*

➤ *The Open dialog box, shown below, appears.*

➤ *From the Open dialog box there are six choices of anatomy content: Dissectible Anatomy, Atlas Anatomy, 3D Anatomy, Clinical Illustrations, Clinical Animations, and Slide Shows. As you work through the lessons in this lab guide, you will be using Dissectible Anatomy, Atlas Anatomy, 3D Anatomy, and Clinical Animations.*

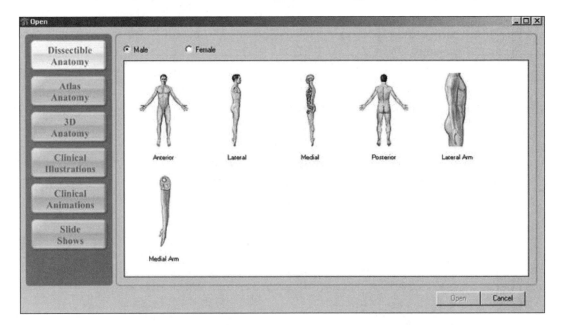

Introduction Exercise 2   Dissectible Anatomy

➤ *Select Dissectible Anatomy from the Open dialog box.*

1. The Open dialog box shows Gender and View options. Select Male and Anterior. Click the Open button. The following appears on your screen:

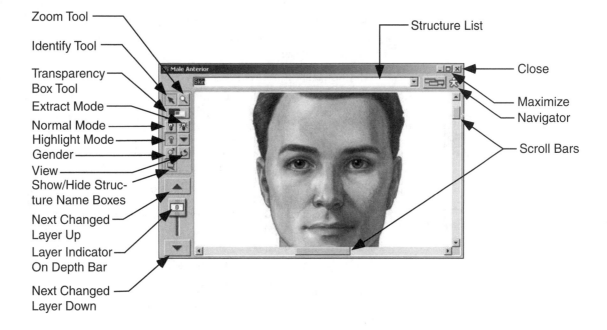

2. The Male Anterior window shows relevant tools used in this lab guide. The functions of these tools are listed below. In Windows, tools can be accessed through the Shortcut menu.

 a. Identify tool—identifies structures. As you click a structure to be identified, its name will appear. After you click a structure, the name of the identified structure then appears in the Structure List. Try it. Click the man's cheek. The word Skin appears.

   b. Zoom tool—increases or decreases magnification. Try it. Zoom out on the face by clicking the Zoom tool and then clicking the face. To Zoom in on the face, select the Zoom tool again and click the face.

   c. Transparency Box tool—restricts dissection to a defined area in the Dissectible Anatomy view and identifies structures at deeper and more superficial layers simultaneously. Try it. Click the Transparency Box tool, and select an area of the body using the cross-hair pointer. A small box with a depth-control slide appears. Move the slider, and the depth in the box changes. Put the Transparency Box tool away by clicking the small box in the upper left corner of the depth-control slide box.

d. Normal—displays all structures in full color.

e. Extract—isolates the selected structure. Try it. Click the lips. Then click the Extract button. The lips appears by themselves with all other structures removed. Click the Normal button to return to a full-color image.

f. Highlight—highlights the selected structure. Try it. Click the lips, then click the Highlight button. Only the lips appear in color. Click the Normal button to return to a full-color image.

g. Gender—changes the gender. Try it. Click and hold the Gender button. A drop-down menu appears with the choice of Male or Female. Drag the cursor down and release the mouse button on Female. The image is now female. Return to the male image.

h. View—changes the view (anterior, posterior, lateral, medial, lateral arm, medial arm). Try it. Click the View button and select Lateral from the drop-down menu. A Male Lateral image appears. Return to the Male Anterior image.

i. Show/Hide Structure Name Boxes—identifies a structure by showing its name in a small pop-up window on the main screen as well as in the Structure List window at the top of the screen. To inactivate the Structure Name Box pop-up window, click the Show/Hide Structure Name Boxes button. Note that the name of the structure remains in the Structure List window. To reactivate the pop-up window, click the Show/Hide Structure Name Boxes button again.

j. Layer Indicator on Depth bar—dissects and restores anatomy layers. Click and drag the Layer Indicator. The number within the Layer Indicator changes, identifying the layer that will be visible when the mouse is released. Adjust the Layer Indicator to 45. Forty-four anatomy layers have been removed, making the facial bones visible.

k. Next Changed Layer (up or down)—removes or adds layers within the region of the image visible in the window. Click the bottom arrow once. The Layer Indicator now shows 46. Click the top arrow once. The Layer Indicator now shows 45.

l. Structure List—displays the name of the item last identified. Click the forehead. The term Frontal bone appears in the Caption bar. Drag the Layer Indicator back to 0.

m. Scroll bars—allow horizontal and vertical positioning of the image.

n. Navigator—navigates to a specific position on the image. Move the cursor over the Navigator icon. A small representation of a male in the anatomical position appears with a box superimposed over the region currently visible on the screen. Move the cursor over the representation, and drag the box to a new anatomical region. The new region is now visible on the screen.

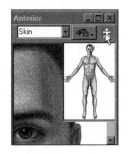

o. Close button—closes the window. Click this box to close the Male Anterior window. The Close button is in the top right corner.

Introduction Exercise 3   Atlas Anatomy

1. Choose Open Content from the File menu and select Atlas Anatomy in the Open dialog box. The following screen appears.

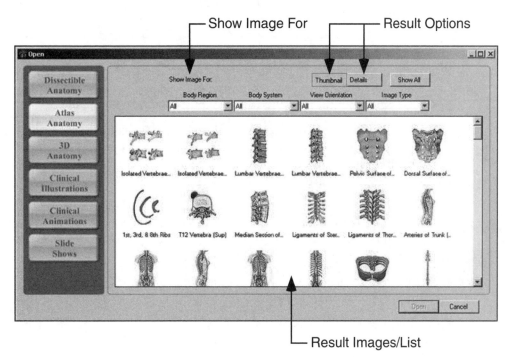

a. Show Image For—allows you to choose images by Body Region, Body System, View Orientation, and Image Type.

b. Result Options—allows the result images to be viewed as thumbnails or a detailed list. Try it. Click the Details button. The Results window shows a detailed list of the images available. Click the Thumbnail button to return to the original window.

c. Result Images/List—results of available images are presented as either thumbnails or a detailed list.

➤ *Select Head and Neck from the Body Region drop-down menu, Integumentary from the Body System drop-down menu, Anterior from the View Orientation drop-down menu, and Illustration from the Image Type drop-down menu. Note as you proceed through these selections that the number of choices is decreased. Select "Surface Anatomy of the Head [Ant]." Click the Open button. Expand the window by clicking the Maximize button.*

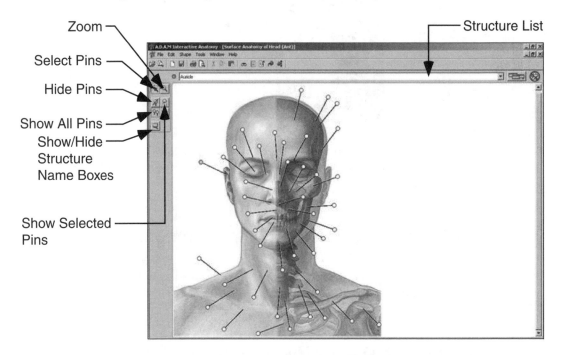

a. Select Pins tool—allows the user to click on the heads of pins. The name of the structure is identified in the Structure List and as a pop-up window on the main screen. Try it. Click on the pin pointing to the right ear. Auricle is identified.

b. Show/Hide Structure Name Boxes—inactivates the Structure Name Box pop-up window by clicking the Show/Hide Structure Name Boxes button. Note that the name of the structure remains in the Structure List window. To reactivate the pop-up window, click the Show/Hide Structure Name Boxes button again.

c. Zoom tool—increases or decreases magnification as in Dissectible Anatomy. In many portions of the lab guide, you are asked to Zoom out on the image so that you can see the entire image. Try it. Zoom out on the face by clicking the Zoom tool and then clicking the face. Now Zoom in again.

d. Hide Pins tool—removes the pins from the image on the screen. Try it. Click the Hide Pins tool. The pins are now removed.

e. Show All Pins tool—shows all pins for any system selected from its menu. Try it. Select Show All Pins from the Options menu. All of the pins are now visible. Now click and hold the down arrow next to the Show All Pins icon. A menu appears. Systems with pinned structures visible within this image are in bold print. Select Integumentary. Only the pins associated with the integumentary system appear.

f. Show One Pin tool—displays only the pin for the selected structure. Try it. Click the Show One Pin button. One pin appears on the image. The pin you selected also appears in the Structure List.

g. Structure List—lists the structures that can be identified in Atlas Anatomy. Try it. Click the Structure List and select *Nose*. A single pin appears on the nose. Now click the Show All Pins icon. Click the Structure List and select Lower Lip. The head of the pin in the Lower Lip turns green.

Introduction Exercise 4 3D Anatomy

➤ *Close the Atlas Anatomy window.*

➤ *Select Open Content from the File menu.*

1. Select 3D Anatomy in the Open dialog box. Eight 3D options appear, as shown below.

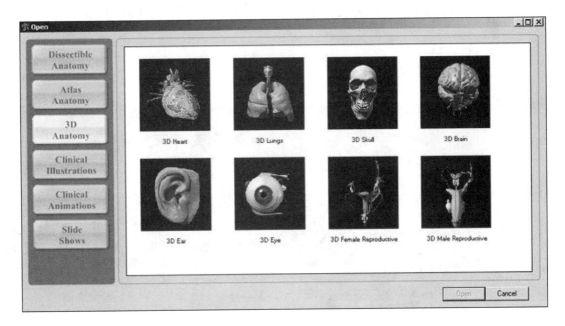

➤ *Select 3D Skull. Click the Open button.*

2. The 3D Skull window, shown below, appears. Relevant tools used in this lab guide are labeled on the diagram. The functions of these tools are listed below the figure.

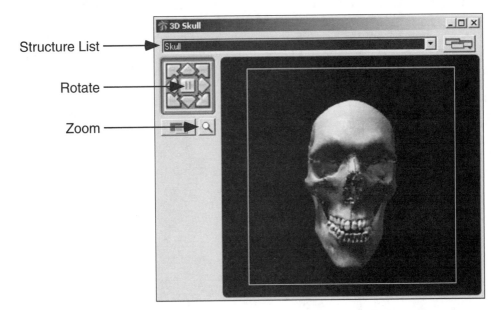

a. Zoom tool—increases or decreases magnification.

b. Rotation controller—rotates the displayed 3D model in selected directions. Click the left arrow. The image of the skull rotates to the left. Click and hold any of the directional arrows. The skull will rotate in the direction chosen.

c. Structure List—lists structures that can be identified in the 3D window. Try it. Click the Structure List and select Sella Turcica—Superior. The skull image rotates, explodes, zooms in, and highlights the sella turcica.

d. Close the 3D Skull window.

Introduction Exercise 5   Clinical Animations

➤ *Select Open Content from the File menu.*

1. Select Clinical Animations. The following Open screen is similar to that of Atlas Anatomy.

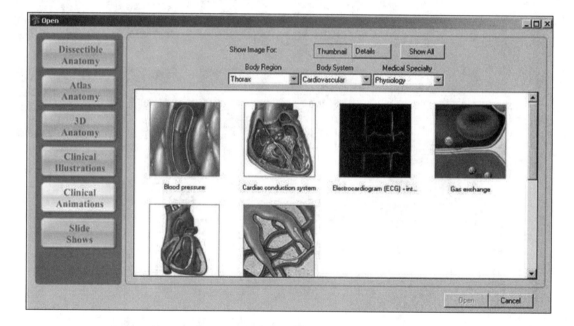

2. Select Thorax from the Body Region drop-down menu, Cardiovascular from the Body System drop-down menu, and Physiology from the Medical Specialty drop-down menu. Select Cardiac Conduction System. Click the Open button. Expand the window by clicking the Maximize button.

3. A Cardiac Conduction System window, like the window on the following page, will open and begin to play. Below the animation window, a text window can be scrolled through for additional information.

4. After the animation finishes, close the Cardiac Conduction System window.

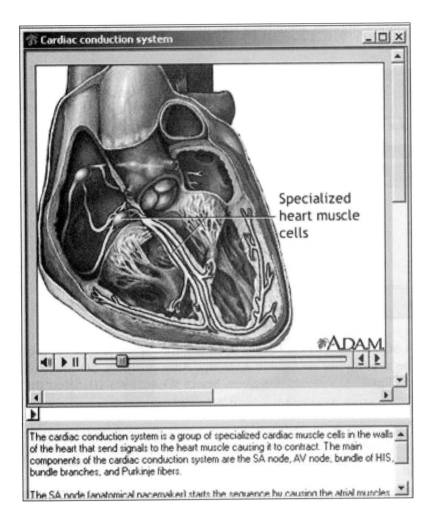

Introduction Exercise 6   Using the Find Feature

➤ *Choose Find from the Tools menu.*

1. Type Skull in the Find box, as shown below. Click the Find button.

➤ *Multiple Functions*

8. Set the Layer Indicator to 13,

   _____

9. with an anterior view,

   _____

10. of the head region

   _____

11. and identify the cartilage forming the tip of the nose.

   _____

12. What are those cartilages called?

   _____

13. Return to full anatomical color.

   _____

14. Isolate/extract these cartilages so that they are the only portion of the image shown on the screen.

   _____

SKELETAL SYSTEM

STUDENT OBJECTIVES

OVERVIEW
- Review the anatomy of the axial and appendicular skeletons.

AXIAL SKELETON

THE SKULL
- Identify the bones of the skull and their surface features as seen in anterior, posterior, superior, inferior, lateral, and medial views.
- Identify specific skull features as seen within an exploded skull.
- Using AIA 3D Anatomy, identify the paranasal sinuses.
- Describe the makeup of the bony orbit.
- Describe the makeup of the nasal septum, including the vomer and ethmoid bones.
- Identify the foramina of the skull.

OTHER BONES OF THE AXIAL SKELETON
- Describe the location and function of the hyoid bone.
- Identify the divisions of the sternum.
- Describe the articulations of the ribs with the sternum anteriorly and the vertebral column posteriorly.
- Name and describe the groups of vertebrae that contribute to the vertebral column.
- Identify intervertebral disks within the vertebral column.

APPENDICULAR SKELETON
- Name the bones of the pectoral girdle, the scapula and clavicle, and identify their major surface markings and articulations.
- Identify the bones of the upper extremity.
- Describe the formation of the bony pelvis, including the two hip bones (composed of the fused ilium, ischium, and pubis) and their articulation with the sacrum and coccyx.
- Identify the bones of the lower extremity.

AXIAL SKELETON

The Skull
Exercise 1.1   Anterior View of the Skull

➤ *Open AIA by double-clicking Start Interactive Anatomy. Choose Dissectible Anatomy.*

➤ *Be sure that Male and the Anterior thumbnail icon are selected. Click Open. Expand the window.*

➤ *Select Find from the Tools menu. A new window appears titled Find. Type "Skull" in the box. Click the Find button.*

➤ *A list of five structures that include the skull appears in the Find Results window. Highlight Skull. Click the Show Results button. Select Male Anterior from the image window that appears. Click Open.*

➤ *An image of the skull appears. The Depth Bar to the left of the image has now moved down, and the Layer Indicator now appears with the number 48 within it. Use the Zoom button to increase the size of the image.*

1. Label the diagram below. Single click on the part of the image that you want to label. The name of the structure appears on both the image and in the title bar above the image. Also, the structure is highlighted. Fill in the names of the structures next to the corresponding letter.

- Frontal bone
- Zygomatic bone
- Maxilla
- Mandible
- Nasal bone

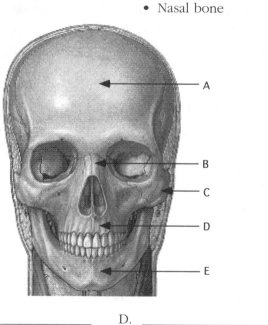

A. _____   D. _____

B. _____   E. _____

C. _____

Exercise 1.2 Lateral View of the Skull

➤ *Select View from the Tool palette. Select Lateral from the drop-down menu.*

➤ *Drag the Depth Bar until the number 187 appears on the Layer Indicator.*

➤ *If necessary, expand the view by clicking the Maximize button in the upper right corner of the window.*

1. Fill in the names of the structures next to the corresponding letter on the following diagram. (Note: Some of the bones identified are labeled in more detail than we are looking for at this time. For instance, upon clicking a specific region of the temporal bone, you may see "Squamous portion of the temporal bone." At this point, we are only going to label the entire bone, or the temporal bone.)

- Temporal bone
- Frontal bone
- Parietal bone
- Occipital bone
- Maxilla

- Mandible
- Zygomatic bone
- Sphenoid bone
- Nasal bone

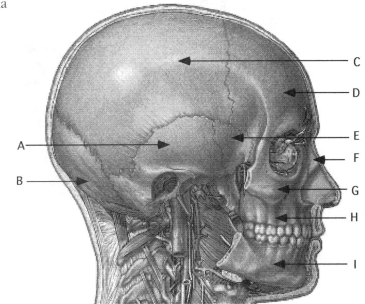

A. _____   F. _____

B. _____   G. _____

C. _____   H. _____

D. _____   I. _____

E. _____

Exercise 1.3 Posterior View of the Skull

➤ *To examine the skull from a posterior view, click the View button, located in the Tool palette, and select Posterior from the drop-down menu that appears.*

➤ *Drag the Depth Bar until the number 186 appears on the Layer Indicator.*

1. Label the following structures on the diagram below (remember to use anatomical position, that is, patient's perspective).

- Occipital bone
- Superior nuchal line
- Inferior nuchal line
- Temporal bone

- Right parietal bone
- Left parietal bone
- External protuberance of occipital bone

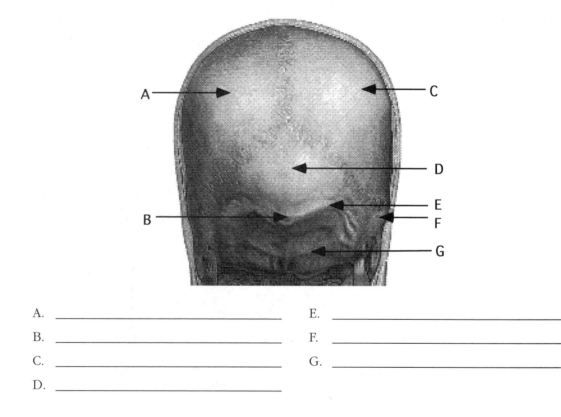

A. _____   E. _____

B. _____   F. _____

C. _____   G. _____

D. _____

Exercise 1.4 Sutures: Lateral View

➤ *Select Lateral from the View button drop-down menu.*

➤ *Adjust the Layer Indicator of the Depth Bar to 186.*

➤ *Adjust the window so that you can see the entire figure.*

1. Label the diagram below.

   • Coronal suture   • Lambdoidal* suture

   • Squamosal suture

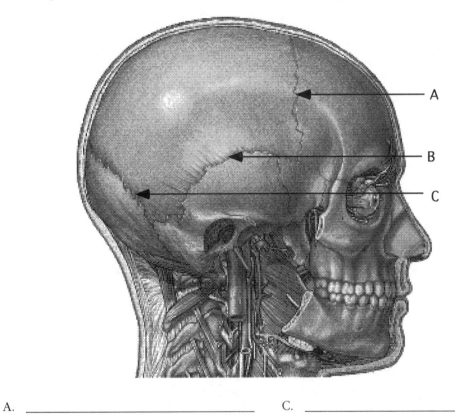

A. _____   C. _____

B. _____

2. What two bones fuse to form the coronal suture?

   _____

3. What two bones fuse to form the lambdoidal suture?

   _____

4. What two bones fuse to form the squamosal suture?

   _____

*Note: AIA identifies this as lambdoid; however, it should be lambdoidal.

Exercise 1.5 Sutures: Posterior View

➤ *Select Posterior from the View button drop-down menu.*

➤ *Adjust the Layer Indicator of the Depth Bar to 186.*

1. Label the following structures on the diagram below.

   • Lambdoidal suture   • Sagittal suture

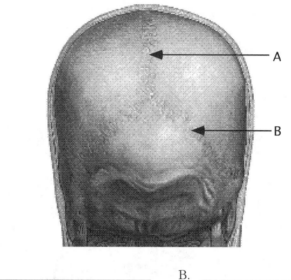

A. _____   B. _____

2. In the posterior view, what bones fuse together to form the sagittal suture?

   _____

3. What bones fuse together to form the lambdoidal suture?

   _____

Exercise 1.6 Frontal Bone Markings

➤ *Choose an Anterior view of the skull and adjust the Layer Indicator to 48.*

1. Label the following structures on the diagram below.

   • Frontal bone   • Supraorbital foramen

   • Supraorbital margin   • Zygomatic process of frontal bone

   • Orbital part of
     the frontal bone

A. _____   D. _____

B. _____   E. _____

C. _____

Exercise 1.7 Sinuses of the Skull

➤ *Choose a Lateral view of the skull and adjust the Layer Indicator to 209.*

1. Label the following structures on the diagram below.

      • Mucosa of sphenoidal sinus   • Mucosa of frontal sinus

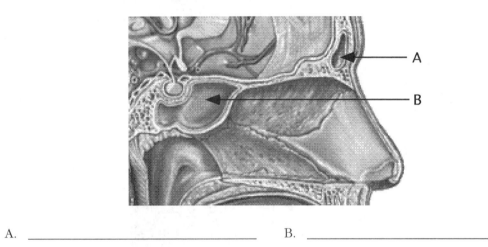

A. _____   B. _____

2. What is the definition of a sinus?

   _____

➤ *To examine all four sinuses, we will use 3D Anatomy. Select Open Content from the File menu. The Open dialog box appears. Click the 3D Anatomy tab and select 3D Skull in the window. Click the Open button. The 3D Skull window appears. Above the skull is a long drop-down menu called the Structure List.*

➤ *Select Tile Vertically from the Windows menu. The lateral view of the skull and the 3D image will be side by side. You may need to close the Find Results window and select Tile Vertically again.*

➤ *Select Sinuses of Skull from the drop-down menu that appears when you click the Structure List. A rotating image of the skull appears with the four sinuses shown in different colors.*

3. What bone sinuses in the skull are represented by the following colors?

   a. Red _____

   b. Blue _____

   c. Yellow _____

   d. Orange _____

Exercise 1.8 Bones of the Orbit and Associated Markings of the Eye

➤ *In the Dissectible Anatomy window, select an Anterior view of the skull and adjust the Layer Indicator to 48.*

➤ *Use the 3D Skull window to identify the structures throughout the rest of this exercise. For instance, click the Structure List at the top of the 3D window and select "Infraorbital foramen" from the list of structures. The image of the skull rotates around and zooms in, and the structure is highlighted.*

1. What color is used to highlight the infraorbital foramen?

_____

2. Using the Male Anterior window and the 3D window, label the diagram below with the following terms.

- Maxilla
- Supraorbital margin
- Lacrimal bone
- Orbital part of greater wing of sphenoid bone
- Ethmoid bone

- Zygomatic bone
- Supraorbital foramen
- Infraorbital foramen
- Lesser wing of sphenoid bone

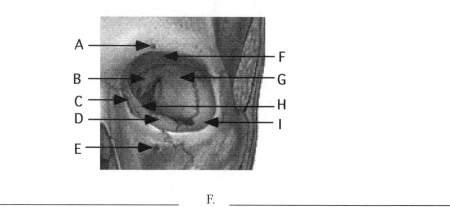

A. _____   F. _____

B. _____   G. _____

C. _____   H. _____

D. _____   I. _____

E. _____

3. What bone contains the supraorbital foramen?

_____

4. What bone contains the infraorbital foramen?

_____

Exercise 1.9 External and Coronal Section Structures of the Nose

1. Use the same view (Male Anterior, Layer Indicator 48) to label the diagram below with the following terms.

    - Maxilla
    - Perpendicular plate of the ethmoid bone

    - Vomer
    - Nasal bone

A. _____   C. _____

B. _____   D. _____

Exercise 1.10 Internal Nose

➢ *Adjust the Layer Indicator to number 49.*

1. Use the Identify tool to label the diagram below with the following terms.

    - Palatine process of the maxilla
    - Crista galli of ethmoid bone
    - Perpendicular plate of ethmoid bone
    - Mucosa of maxillary sinus

    - Perpendicular plate of vomer
    - Middle nasal concha
    - Inferior nasal concha

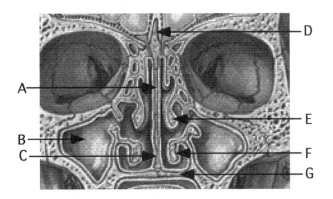

A. _____   E. _____

B. _____   F. _____

C. _____   G. _____

D. _____

2. Label the following structures as seen in the 3D window.

   • Middle nasal concha   • Inferior nasal concha

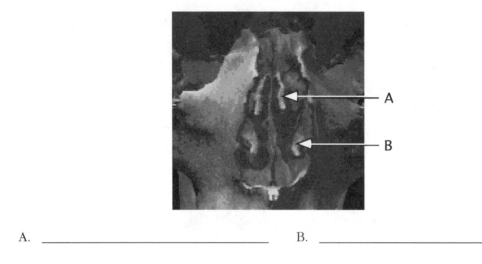

A. _____   B. _____

➤ *The Lateral view of the nasal septum further illustrates the makeup of the nasal septum.*
   Choose a Lateral view of the skull and adjust the Layer Indicator to 209.

3. Use the Identify tool to label the diagram below with the following terms.

   • Perpendicular plate of ethmoid bone   • Vomer

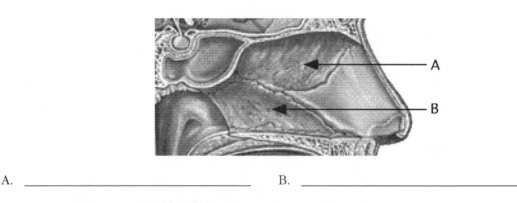

A. _____   B. _____

4. The septum of the nose is divided into the superior and inferior portions of the perpendicular plate. What bone forms the superior portion of the perpendicular plate?

   _____

5. Describe the formation of the inferior portion of the perpendicular plate.

   _____

Exercise 1.11 Temporal Bone, Zygomatic Bone, and Zygomatic Arch

➤ *In the Lateral view, adjust the Layer Indicator to 118.*

1. Label the diagram below. For this diagram (and the diagrams that follow) you are not given the terms to be labeled. Determine the proper labels by clicking on the structure you want to label.

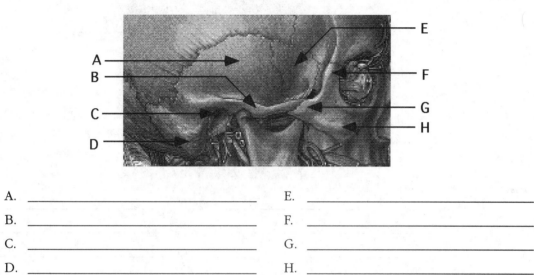

A. _____ E. _____

B. _____ F. _____

C. _____ G. _____

D. _____ H. _____

2. What two structures make up the zygomatic arch?

_____

➤ *To see other structures associated with the temporal bone, it is necessary to remove the zygomatic arch and a portion of the mandible. To do this, adjust the Layer Indicator to 129.*

3. Label the following diagram.

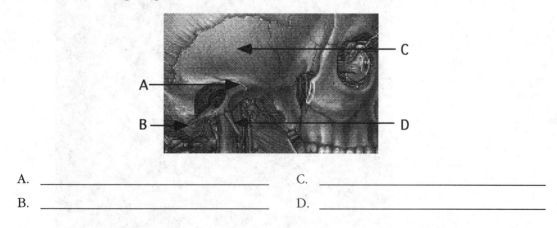

A. _____ C. _____

B. _____ D. _____

Exercise 1.12 Mandible

➤ *In the Lateral view, adjust the Layer Indicator to 118.*

1. Use the Dissectible Anatomy window to label the following diagram. (Note: You will need to use the 3D window to locate the mental foramen as it is not identified in Dissectible Anatomy in this view.)

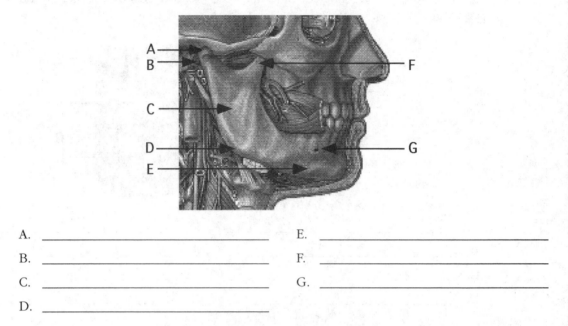

A. _____   E. _____

B. _____   F. _____

C. _____   G. _____

D. _____

2. Use the 3D Anatomy window to identify and label the structures in the following diagram.

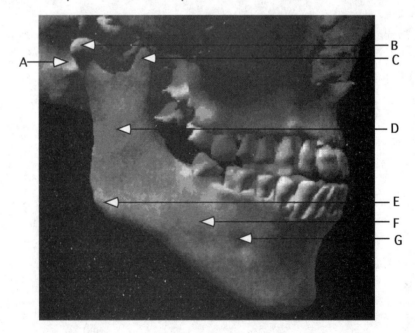

A. _____   E. _____

B. _____   F. _____

C. _____   G. _____

D. _____

3. The temporomandibular joint is the region where the mandible articulates with the temporal bone. What two specific portions of these bones make up the temporo-mandibular joint?

_____

Exercise 1.13 Sphenoid and Ethmoid Bones

➤ *We have already seen that the sphenoid bone contributes to the inner orbital surface as well as to a portion of the lateral skull. To understand why the sphenoid bone is considered the "keystone of the cranium," we will examine the sphenoid bone in the 3D window. Select "Sphenoid bone—Isolated" from the drop-down menu in the Structure List.*

➤ *If you can see only a portion of the sphenoid bone, you will need to use the Zoom tool to reduce the image. (Click the Zoom button in the 3D window and move the cursor over the image. If you see a circle with a minus sign in it, click the 3D window once. If you see a circle with a plus sign in it, you are already in the reduced size.) Do the same thing for the ethmoid bone. (Note: There are two "Ethmoid bone" designations in the Structure List. Select "Ethmoid bone—Isolated.")*

1. Several specific structures of the sphenoid and ethmoid bones can be seen from a superior view with the skull cap removed. To see these structures, use the 3D window. Select each of the following terms from the drop-down menu on the Structure List so that you can label the diagram below.

    • Greater wing of sphenoid bone

    • Lesser wing of sphenoid bone

    • Sella turcica—Superior

    • Crista galli of ethmoid bone

    • Cribriform plate of ethmoid bone

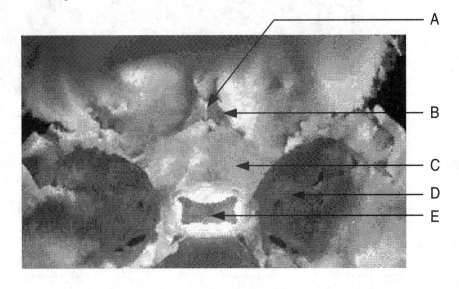

A. _____   D. _____

B. _____   E. _____

C. _____

2. What endocrine structure lies in the sella turcica of the sphenoid?

_____

3. What cranial nerve passes through the foramina in the cribriform plate of the ethmoid?

_____

Exercise 1.14 Inferior View of the Skull: Foramina

1. Now examine various foramina that are visible in an inferior view of the skull using the 3D window. Select the following terms from the Structure List to label the diagram below.

- Foramen magnum
- Jugular foramen—Inferior
- Foramen lacerum—Inferior
- Foramen ovale of sphenoid bone—Inferior
- Foramen spinosum—Inferior

- Incisive fossa
- Carotid canal
- Greater palatine canal
- Stylomastoid foramen

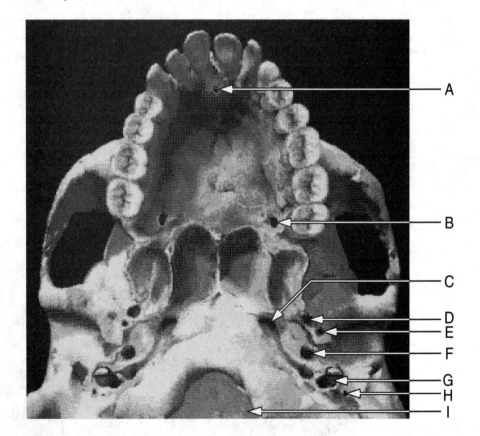

A. _____	F. _____
B. _____	G. _____
C. _____	H. _____
D. _____	I. _____
E. _____	

Exercise 1.15 Superior View of the Base of the Skull: Foramina

1. Now examine various foramina that are visible in a superior view of the cranial floor using the 3D window. Select the following terms from the Structure List and label the diagram below.

- Jugular foramen—Superior
- Foramen lacerum—Superior
- Optic canal—Superior
- Foramen ovale of sphenoid bone—Superior

- Foramen spinosum—Superior
- Hypoglossal canal
- Foramen rotundum—Superior

(Note: You should also label the following terms, although they cannot be directly identified from this superior view.)

- Carotid canal
- Foramen magnum

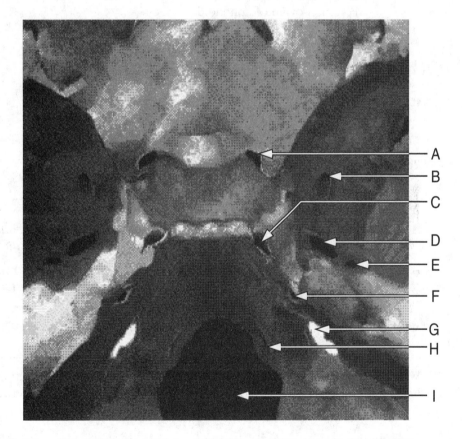

A. _____ F. _____

B. _____ G. _____

C. _____ H. _____

D. _____ I. _____

E. _____

➢ *Match the following foramina with the structures that pass through them.*

2. _____ Carotid canal   A. Hypoglossal nerve

3. _____ Foramen lacerum   B. Nasopalatine nerves

4. _____ Foramen magnum   C. Internal carotid artery

5. _____ Stylomastoid foramen   D. Glossopharyngeal, vagus, and accessory nerves

6. _____ Foramen rotundum   E. Greater palatine vessel and nerves

7. _____ Foramen spinosum   F. Maxillary nerve (CN V2: Trigeminal)

8. _____ Greater palatine canal   G. Optical nerve

9. _____ Hypoglossal canal   H. Spinal cord

10. _____ Incisive fossa   I. Facial nerve

11. _____ Jugular foramen   J. Meningeal branch of (CN V3: Trigeminal)

12. _____ Optic canal   K. Mandibular nerve (CN V3: Trigeminal)

13. _____ Foramen ovale of sphenoid

Other Bones of the Axial Skeleton
Exercise 1.16 Hyoid Bone

➢ *Close 3D Anatomy.*

➢ *To find the hyoid bone, we will use the Find feature of AIA.*

➢ *Select Find from the Tools menu. A Find window appears. Type "Hyoid bone" in the box. Click Find.*

➢ *A list of five structures appears in the Find Results box. Select Hyoid bone. Click the Show Results In button and select Male and the Anterior thumbnail icon. Click Open.*

➢ *A figure of the neck region appears with the hyoid bone highlighted. You will also note that the Depth Bar to the left of the image has now moved down and the Layer Indicator now appears with the number 254.*

1. What color does AIA use to highlight the hyoid bone?

_____

2. Where is the hyoid bone located in relation to the larynx? (Note: You can move the image up and center the hyoid bone on your screen by using the vertical and horizontal scroll bars.)

_____

3. In terms of articulation with another bone, how is the hyoid bone unique in the axial skeleton?

_____

4. What is the anatomical function of the hyoid bone?

_____

5. What action does the sternohyoid muscle have on the hyoid bone?

_____

Exercise 1.17 Sternum and Ribs

➤ *Select Open Content from the File menu and choose Dissectible Anatomy. Select Anterior and then click Open. Maximize the image and set the Layer Indicator to 157.*

➤ *Select Open Content from the File Menu and choose Atlas Anatomy. Below the Body Region drop-down menu, select Thorax. Beneath the Image Type drop-down menu, select Cadaver Photograph. Click on the "Dissection of Thorax (Ant)" thumbnail icon. Click Open. Be sure that Show All Pins is selected.*

➤ *Select Tile Vertically from the Window menu. Center the image in the respective windows.*

1. Fill in the name of the structure next to the corresponding letter in the space below the diagram.

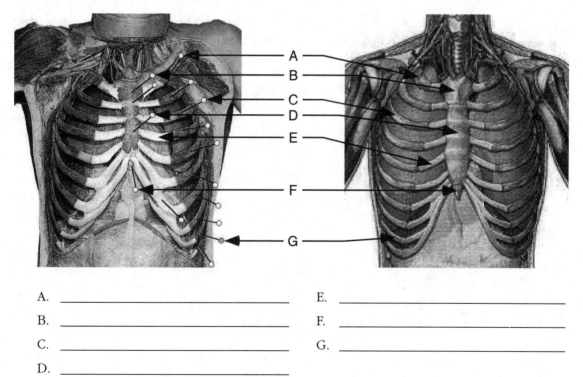

A. _____   E. _____

B. _____   F. _____

C. _____   G. _____

D. _____

2. Superior to the manubrium (along the midline) is a white-ringed, tubelike structure. What is it?

_____

3. What is the principal anatomical function of the sternum?

_____

4. By what name is the junction between the manubrium and sternal body known?

_____

5. The most inferior portion of the sternum is the xiphoid process. What does the term _xiphoid_ mean, literally, and to what sternal feature does it apply?

_____

6. What is the clinical significance of the sternal angle?

_____

7. What is a sternal puncture?

_____

8. What bone of the appendicular skeleton articulates with the sternum, and, in terms of upper limb stability, what role does the identified bone play?

_____

9. Why are the first seven pairs of ribs known as true ribs?

_____

10. Why are ribs 8 through 12 known as false ribs?

_____

11. Why are ribs 11 and 12 known as floating ribs?

_____

Exercise 1.18 Vertebral Column

➤ *Close the "Dissection of Thorax (Ant)" window.*

➤ *Select Posterior from the View button drop-down menu.*

➤ *Adjust the Layer Indicator to 176.*

➤ *To view the vertebral column in full color, click the Normal button in the Tool palette.*

➤ *Click the first vertebra immediately beneath the skull. It will either be identified as the "Posterior tubercle of C1 vertebra {atlas}" or "Posterior arch of C1 vertebra {atlas}." In either case, you have identified the first cervical vertebra, whose common name is the atlas.*

1. How many cervical (C) vertebrae are there? (How many vertebrae are identified as " . . . of C _____"?)

   _____

2. How many thoracic (T) vertebrae are there?

   _____

3. What flat bones articulate with the thoracic vertebrae?

   _____

4. How many lumbar (L) vertebrae are there?

   _____

5. Describe the location and bony makeup of the sacrum.

   _____

6. What is the coccyx?

   _____

➤ *To examine the vertebral column from an anterior view, select Anterior from the View button drop-down menu.*

➤ *Adjust the Layer Indicator to 329.*

7. Label the following structures on the diagram below:

   - Thoracic vertebrae
   - Intervertebral disc
   - Lumbar vertebrae
   - Cervical vertebrae
   - Rib
   - Coccyx
   - Sacrum

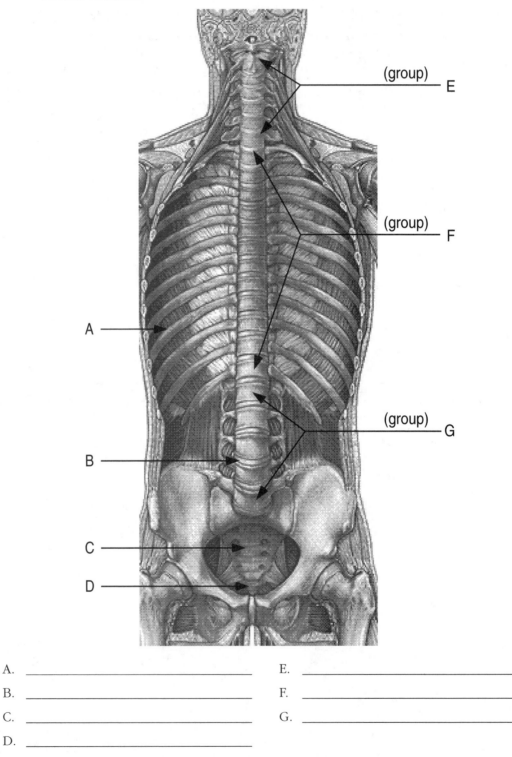

A. _____   E. _____

B. _____   F. _____

C. _____   G. _____

D. _____

8. What is the common name of the second cervical vertebra, and how is this name related to its function?

_____

9. The second cervical vertebra contains a specialized projection known as the dens or odontoid process. What is the derivation of this projection's name?

_____

10. What is the function of the dens?

_____

11. Why is cervical vertebra number 7 known as the vertebra prominens?

_____

12. Intervertebral discs consist of fibrocartilage. What does this imply concerning disc function?

_____

13. Why are the thoracic and sacral curves known as primary curvatures of the spine?

_____

14. Why are the cervical and lumbar curves known as secondary curvatures of the spine?

_____

15. What is the function of the body of a vertebra?

_____

16. In what important respect are the transverse processes of the cervical vertebrae different from other vertebral transverse processes?

_____

17. A special posterior feature of the sacrum is the median sacral crest. What vertebral feature does this structure represent, and how is this sacral feature different from the same structure on lumbar vertebrae?

_____

Exercise 1.19 Lumbar Vertebrae and Hip Region

1. Use the same image to label the structures below on the following diagram.
   - Body of L5 vertebra
   - Body of L1 vertebra
   - Body of L4 vertebra
   - Sacrum
   - Intervertebral disc
   - Coccyx

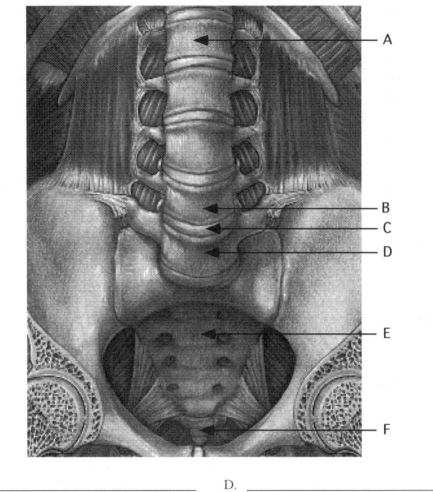

A. _____   D. _____

B. _____   E. _____

C. _____   F. _____

Exercise 1.20 Vertebra Structure: Posterior View

➢ *Select a Posterior view using the image View button at the left of the screen. Return to full color if necessary.*

➢ *Adjust the Layer Indicator to 176.*

➢ *Locate the second thoracic vertebra below the junction of the cervical and thoracic vertebrae. Zoom in on this region if necessary.*

1. Label the vertebral structures below on the following diagram.

    • Spinous process of T2 vertebra   • Transverse process of T2 vertebra

    • Inferior articular process of T2 vertebra   • Lamina of T2 vertebra

    • Superior articular process of T2 vertebra

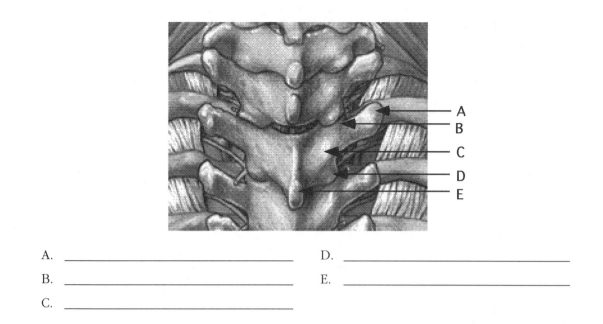

A. _____   D. _____

B. _____   E. _____

C. _____

Exercise 1.21 Vertebra Structure: Lateral View

➤ *Select a Lateral view using the image View button at the left of the screen. Return to full color if necessary.*

➤ *Adjust the Layer Indicator to 275.*

1. Locate the second thoracic vertebra and label the following vertebral structures on the diagram below.

   - Spinous process of T2 vertebra
   - Body of T2 vertebra
   - Facet for tubercle of rib of T2 vertebra
   - Pedicle of T2 vertebra
   - Inferior articular process of T2 vertebra
   - Intervertebral disc
   - Superior articular process of T2 vertebra

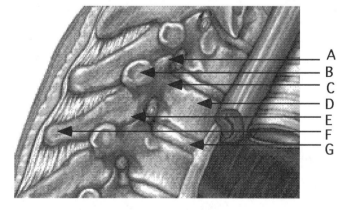

A. _____   E. _____

B. _____   F. _____

C. _____   G. _____

D. _____

APPENDICULAR SKELETON

Pectoral Girdle
Exercise 1.22 Clavicle

➤ *Select Anterior from the View button drop-down menu.*

➤ *Select Find from the Tools menu. A Find window appears. Type "Clavicle" in the box. Click Find.*

➤ *A list of four structures appears in the Find Results box. Select Clavicle. Now click the Show Results In button. Select Male Anterior. Click on Open.*

➤ *A figure of the anterior neck region appears with the clavicle highlighted. You will also note that the Layer Indicator in the Depth Bar now appears with 56 in it.*

1. Adjust the image on your screen so that it matches the diagram below, and label the three regions of the clavicle shown in the figure.

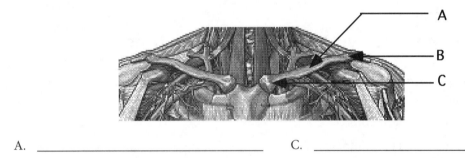

A. _____   C. _____

B. _____

2. With what specific structure does the medial end of the clavicle articulate?

_____

3. With what specific structure does the lateral end of the clavicle articulate?

_____

Beyond A I A

4. The clavicle can be palpated through its entire length and is said to follow a sinuous curve. What is meant by *sinuous curve*?

_____

5. Why might falling on one's outstretched arm result in a fractured clavicle?

_____

6. Fractures of the clavicle, especially in children, are known as greenstick fractures. Describe this type of fracture.

_____

Exercise 1.23 Scapula

➢ *Select Posterior from the View button drop-down menu.*

➢ *Adjust the Layer Indicator of the Depth Bar to 72.*

➢ *Click the Normal Mode button.*

➢ *Adjust the image so that the right posterior shoulder region appears on your screen.*

1. The pink structure covering the head of the humerus where it articulates with the scapula is the articulating cartilage. What is the function of this structure?

   _____

➢ *Select Open Content from the File menu and choose Dissectible Anatomy. Select Lateral in the View menu. Click Open. A new window, Male Lateral, appears. Adjust the image on your screen so that a lateral view of the shoulder is seen. Adjust the Layer Indicator of the Depth Bar to 41.*

➢ *Select Tile Vertically from the Windows menu. All the open windows in the program appear tiled side by side. Close all the windows except the lateral and posterior views of the scapula. Select Tile Vertically again. Now only the lateral and posterior views of the scapula are side by side.*

2. Label the two views that follow.

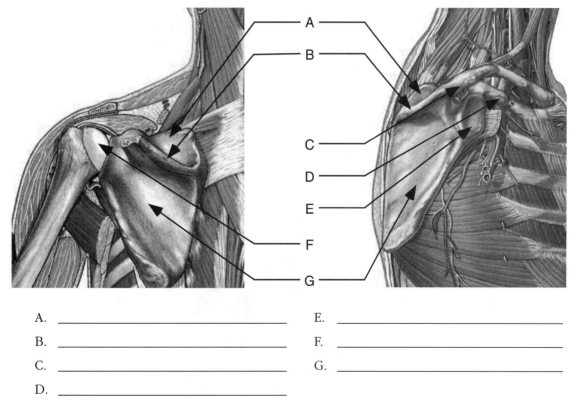

A. _____	E. _____
B. _____	F. _____
C. _____	G. _____
D. _____	

Beyond
A I A   3. Where is the supraspinatus muscle located?

_____

4. Where is the infraspinatus muscle located?

_____

5. Why is the acromion process known as the "summit of the shoulder"?

_____

6. What is another name for the medial border of the scapula, and why are both names appropriate?

_____

➤ *Close the Male Posterior window.*

Bones of the Upper Extremity
Exercise 1.24 Humerus

➤ *Select Anterior from the View button drop-down menu.*

➤ *Select Find from the Tools menu. Type "Humerus" in the box. Click Find.*

➤ *Select the term Humerus. Now click the Show Results In button. Select Male Anterior. Click Open. The Layer Indicator should show 329.*

➤ *Adjust the image so that it matches the diagram below.*

1. Use the Identify tool to label the diagram.

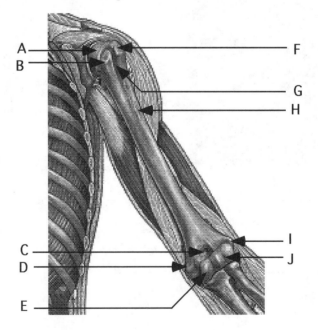

A. _____   F. _____

B. _____   G. _____

C. _____   H*. _____

D. _____   I. _____

E. _____   J. _____

*Note: It may take a few tries to find the name of this tuberosity.

2. What are the names of the two bones that articulate with the distal humerus?

_____

3. What is the capitulum?

_____

4. What is the trochlea of the humerus?

_____

5. What is the name of the bony structure on the scapula that articulates with the head of the humerus, and what does this structure's name mean, literally?

_____

6. What is the name of the groove found between the greater and lesser tubercles of the humerus?

_____

Beyond AIA

7. How is the long head of the biceps brachii related to the intertubercular (bicipital) groove?

_____

8. What is the function of the olecranon fossa?

_____

Exercise 1.25 Forearm, Wrist, Hand, and Fingers

➤ *Using the Navigator tool, adjust the image so that the right forearm and hand occupy the screen.*

1. Label the diagram below with the following general terms. (Note: The AIA program may identify these bones in more specific terms.)

- Radius
- Head of the radius
- Styloid process of radius
- Coronoid process of ulna
- Head of the ulna
- Proximal phalanx
- Middle phalanx

- Humerus
- Radial tuberosity
- Ulna
- Metacarpals
- Carpals
- Distal phalanx

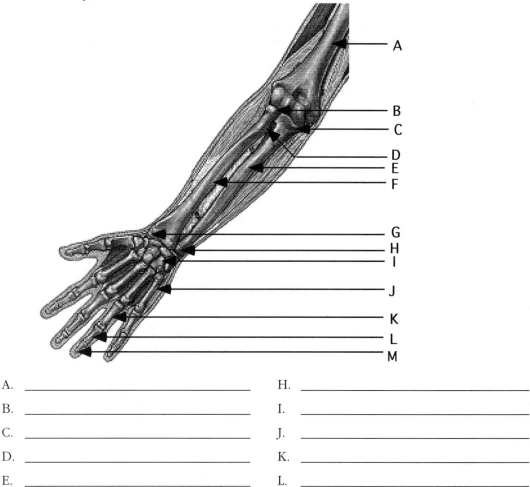

A.	_____	H.	_____
B.	_____	I.	_____
C.	_____	J.	_____
D.	_____	K.	_____
E.	_____	L.	_____
F.	_____	M.	_____
G.	_____		

2. The thumb has only two phalanges. What are they?

_____

3. How many bones make up the wrist, hands, and fingers of the right hand?

_____

4. Zoom in on the carpals. How many bones of the carpus can be identified?

_____

5. Describe the position of the radius with respect to the ulna during forearm supination.

_____

6. Describe the position of the radius with respect to the ulna during forearm pronation.

_____

Pelvic Girdle
Exercise 1.26 Hip Bones

➤ *Using the Navigator tool, adjust the image on your screen so that it matches the diagram below.*

1. Label the diagram using the Identify tool.

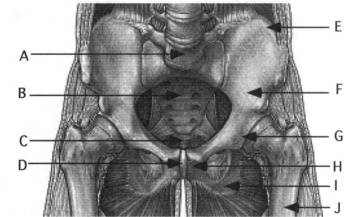

A. _____   F. _____

B. _____   G. _____

C. _____   H. _____

D. _____   I. _____

E. _____   J. _____

Beyond A I A

2. What does the "L" signify in "Vertebral body of L5"?

_____

3. What is the acetabulum?

_____

4. What three bones fuse to form the hip bone?

_____

Bones of the Lower Extremity
Exercise 1.27 Femur

➢ *Adjust the image so that the left anterior femur is visible. Set the Layer Indicator to 329.*

➢ *Select Open Content from the File menu and choose Dissectible Anatomy. Select Posterior, and click Open. A new window, Male Posterior, appears. Adjust the image in the Male Posterior window so that the upper left leg is seen. Adjust the Layer Indicator to 185. Zoom out and adjust the image so that the left posterior femur is centered in the window.*

➢ *Select Tile Vertically from the Window menu. The anterior and posterior views of the femur will be side by side. Adjust the images to match the diagram below.*

1. Fill in the names of the structures next to their corresponding letters in the spaces below the diagram.

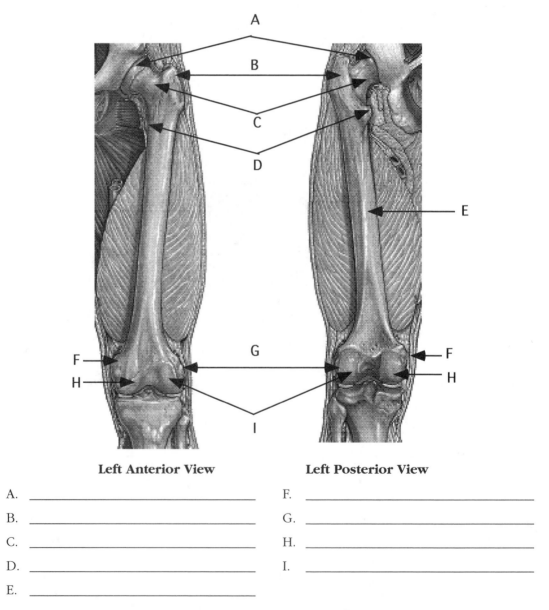

Left Anterior View   **Left Posterior View**

A. _____   F. _____

B. _____   G. _____

C. _____   H. _____

D. _____   I. _____

E. _____

2. What is significant about the length of the femur in relation to the other bones of the body?

_____

3. With which lower leg bone does the femur articulate?

_____

Exercise 1.28　Bones of the Lower Leg

➤ *Close the Male Posterior window.*

➤ *Adjust the Male Anterior window so that it is centered on the right anterior leg as shown below.*

1. Label the diagram with the following terms. (Note: AIA may identify these bones in more specific terms.)

- Tarsals
- Metatarsals
- Proximal phalanx
- Distal phalanx
- Middle phalanx
- Fibula
- Tuberosity of tibia

- Medial malleolus of tibia
- Lateral malleolus of fibula
- Medial condyle of tibia
- Lateral condyle of tibia
- Femur
- Tibia
- Talus bone

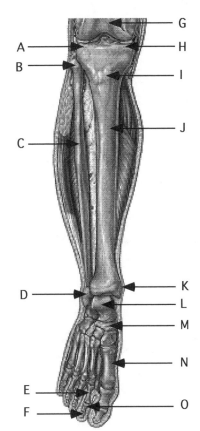

A. _____   I. _____

B. _____   J. _____

C. _____   K. _____

D. _____   L. _____

E. _____   M. _____

F. _____   N. _____

G. _____   O. _____

H. _____

2. What is the anatomic name given to the bone that covers over the articulation between the thigh and the leg, and to what special bone group does this bone belong?

_____

3. What is the name of the distal expansion of the tibia (often referred to as the internal ankle)?

_____

4. What is the name of the distal expansion of the fibula (often referred to as the external ankle)?

_____

2

MUSCULAR SYSTEM

STUDENT OBJECTIVES

OVERVIEW

- Review the anatomy of the major superficial and deep muscles of the body.

SUPERFICIAL AND DEEP MUSCLES OF THE BODY

- Identify and describe the major muscles of the head (anterior, posterior, and lateral), neck (anterior, posterior, and lateral), shoulder (anterior and posterior), chest (anterior), back (posterior), abdomen (anterior), arm and forearm (anterior and posterior), thigh (anterior), buttock and thigh (posterior), and leg (anterior, posterior, and lateral).

- Discover various methods employed to name skeletal muscles.

- Describe the action of some specific skeletal muscles.

- Determine the proximal attachment (PA) (origin) and distal attachment (DA) (insertion) of various muscles.

SUPERFICIAL AND DEEP MUSCLES OF THE UPPER BODY

Exercise 2.1 Anterior Muscles of the Head, Neck, Shoulder, and Chest

➤ *Open AIA by double-clicking Start Interactive Anatomy. Select Dissectible Anatomy.*

➤ *Select Male and the Anterior thumbnail icon. Click Open.*

➤ *Expand the image. Set the Layer Indicator to number 8.*

➤ *Zoom out on the image.*

Beyond AIA

1. What is the derivation of *platysma*?

   _____

2. Adjust the Layer Indicator to 10. Identify the composite sphincter muscle encircling each eye.

   _____

Beyond AIA

3. What is the meaning of the two parts of this muscle's name?

   a. _____

   b. _____

4. Use the Identify tool to label the diagram below.

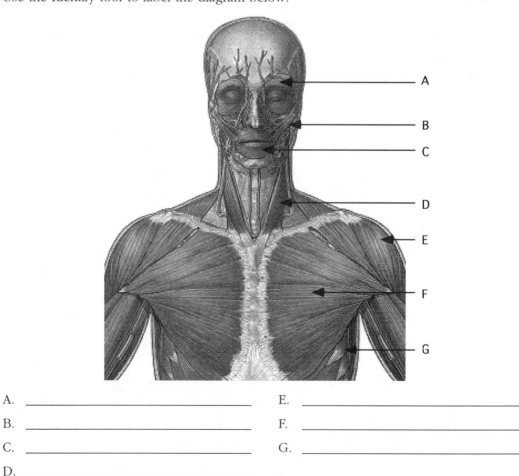

A. _____   E. _____

B. _____   F. _____

C. _____   G. _____

D. _____

Beyond
A I A 5. What is the derivation of *sternocleidomastoid*?

_____

6. What does *serratus anterior* mean?

_____

7. What does the name *serratus anterior* tell you about this muscle relative to its shape and location?

_____

➤ *Select Lateral from the View button drop-down menu. Adjust the Layer Indicator to 10.*

8. Use the Identify tool to label the diagram below.

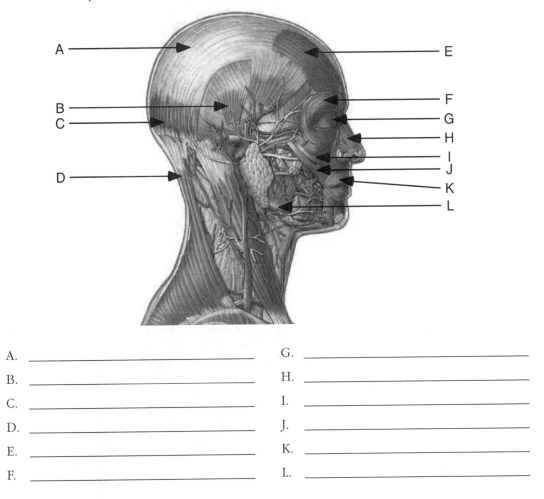

A. _____	G. _____
B. _____	H. _____
C. _____	I. _____
D. _____	J. _____
E. _____	K. _____
F. _____	L. _____

Beyond
A I A 9. The aponeurosis uniting the two portions of the epicranius muscle (frontalis and occipitalis muscle) is called the "galea aponeurotica." The term "galea" means "helmet." Why is this an appropriate name?

_____

10. The masseter muscle is a muscle of mastication. What does this imply?

_____

11. What is the function of the internal jugular vein?

_____

➤ *Adjust the Layer Indicator to 115.*

12. Use the Identify tool to label the diagram below.

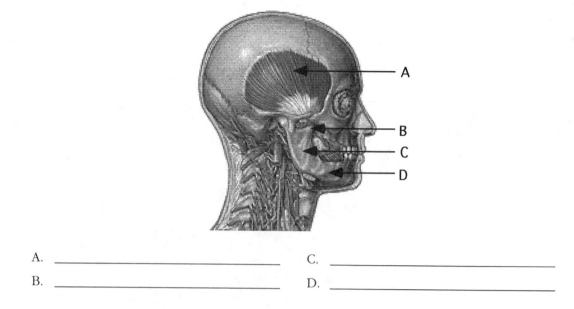

A. _____   C. _____

B. _____   D. _____

➤ *Click on File and select Open Content. Select Atlas Anatomy. From the Region drop-down menu, select Head and Neck. From the System drop-down menu, select Muscular. From the View drop-down menu, select Lateral. From Type, select Cadaver Photograph. Click on the "Muscles of Facial . . . " thumbnail icon and click Open.*

13. Identify the pinned structures in the image below.

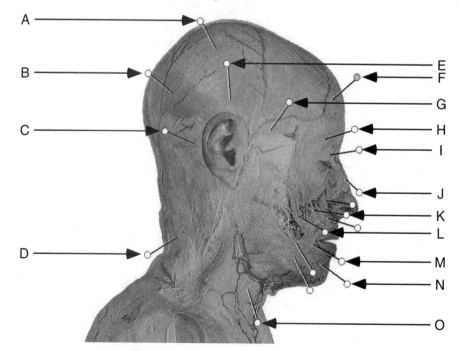

A. _____   I. _____

B. _____   J. _____

C. _____   K. _____

D. _____   L. _____

E. _____   M. _____

F. _____   N. _____

G. _____   O. _____

H. _____

Beyond
A I A

14. The insertion of the temporalis muscle is the coronoid process of the mandible. What is the action of this muscle?

_____

Exercise 2.2 Deep Anterior Muscles of the Shoulder

➤ *Using Dissectible Anatomy, select the Anterior thumbnail icon. Click open.*

➤ *Adjust the Layer Indicator to 53. (In this view, pectoralis major and deltoid muscles have been removed.)*

1. Use the Identify tool to label the diagram below.

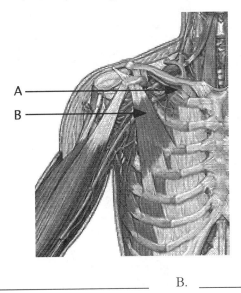

A. _____   B. _____

Beyond
A I A

2. What is the derivation of *subclavius?*

_____

(Note: Muscle B helps stabilize the scapula.)

Exercise 2.3 Posterior Muscles of the Neck, Shoulder, and Back

➢ *Select Posterior from the View button drop-down menu. Adjust the Layer Indicator to 9.*

1. Use the Identify tool to label the diagram below.

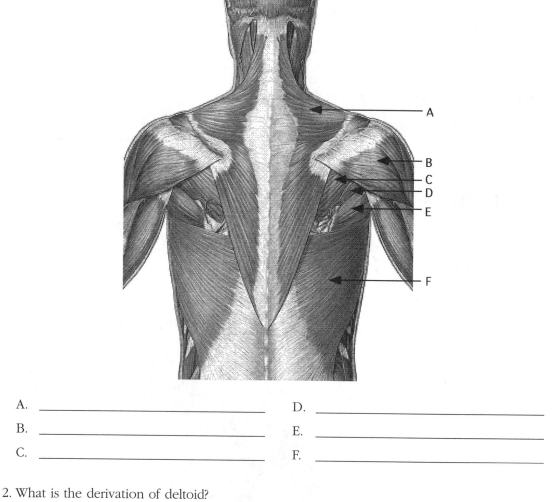

A. _____   D. _____

B. _____   E. _____

C. _____   F. _____

Beyond
A I A

2. What is the derivation of deltoid?

_____

3. What is the derivation of *latissimus dorsi*?

_____

4. What is the tissue composition and significance of the *thoracolumbar fascia*?

_____

Exercise 2.4 Deep Posterior Muscles of the Neck, Shoulder, and Back

➤ *Select Posterior from the View button drop-down menu.*

➤ *Adjust the Layer Indicator to 12. (Both trapezius and latissimus dorsi have been removed in this view.)*

1. Use the Identify tool to label the diagram below.

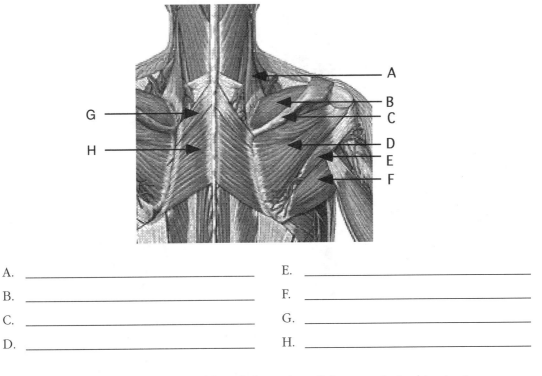

A. _____	E. _____
B. _____	F. _____
C. _____	G. _____
D. _____	H. _____

2. What muscle lies just superior to (above) the spine of the scapula in this view?

_____

3. What muscle lies just inferior to (below) the spine of the scapula in this view?

_____

(Note: Muscles B, D, and E identified in the view above are three of the four muscles that form the rotator cuff. These muscles' tendons "S.I.T." on the humeral head and help stabilize the shoulder through a full range of movement. The fourth rotator cuff muscle, subscapularis, cannot be seen in this view because it lies in front of the scapula.)

4. The levator scapulae is the only neck muscle whose name indicates its action. What is the action of this muscle?

_____

Exercise 2.5 Muscles of the Posterior Arm and Forearm

➢ *Select a Posterior view from the side toolbar.*

➢ *Set the Layer Indicator to 9.*

➢ *Adjust the image on your screen so that the entire posterior portions of the right arm, forearm, and wrist are in view.*

1. Label the muscles and related structures on the diagram below.

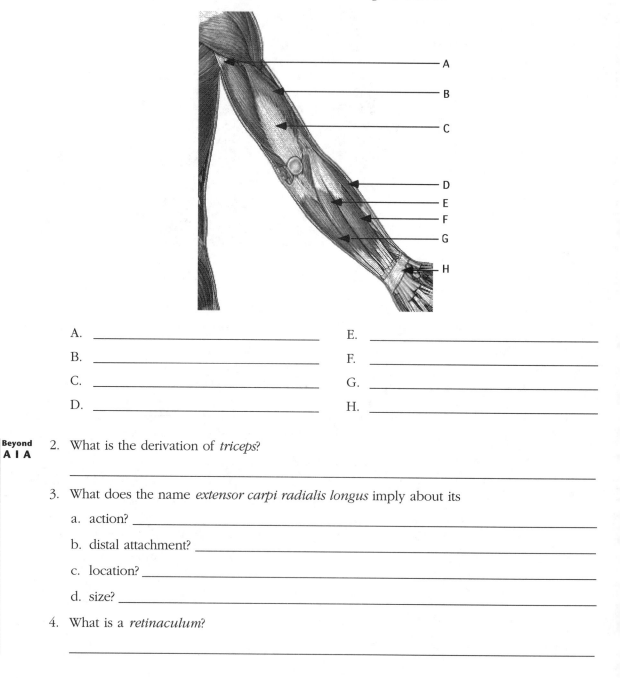

A. _____   E. _____

B. _____   F. _____

C. _____   G. _____

D. _____   H. _____

Beyond A I A

2. What is the derivation of *triceps?*

_____

3. What does the name *extensor carpi radialis longus* imply about its

a. action? _____

b. distal attachment? _____

c. location? _____

d. size? _____

4. What is a *retinaculum?*

_____

Exercise 2.6 Muscles of the Anterior Arm and Forearm

➤ *Select Open Content from the File Menu. Choose Dissectible Anatomy and the Anterior thumbnail icon. Click Open.*

➤ *Enlarge the image and zoom in on the right anterior forearm region. Set the Layer Indicator to 81.*

➤ *Select Open Content from the File Menu. Choose Atlas Anatomy and select the following from the indicated drop-down menus:*

Region: Upper Limb   *View: Anterior*

System: Muscular   *Type: Cadaver Photograph*

➤ *Click on the "Anterior Arm" thumbnail icon. Click Open.*

➤ *Select Tile Vertically from the Window menu. Adjust the images to match the illustrations that follow.*

1. Label the diagram below.

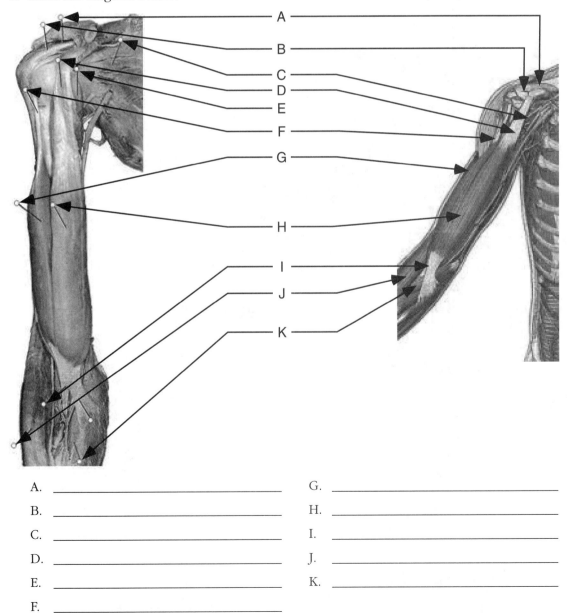

A. _____   G. _____

B. _____   H. _____

C. _____   I. _____

D. _____   J. _____

E. _____   K. _____

F. _____

2. What do the terms *biceps* and *brachii* mean?

a. _____

b. _____

Exercise 2.7 Deep Muscles of the Anterior Arm

➤ *Using Dissectible Anatomy, select Anterior from the View button drop-down menu. Adjust the Layer Indicator to 87.*

1. Observe the left arm, forearm, and wrist. Label the muscles and related structures in the following diagram.

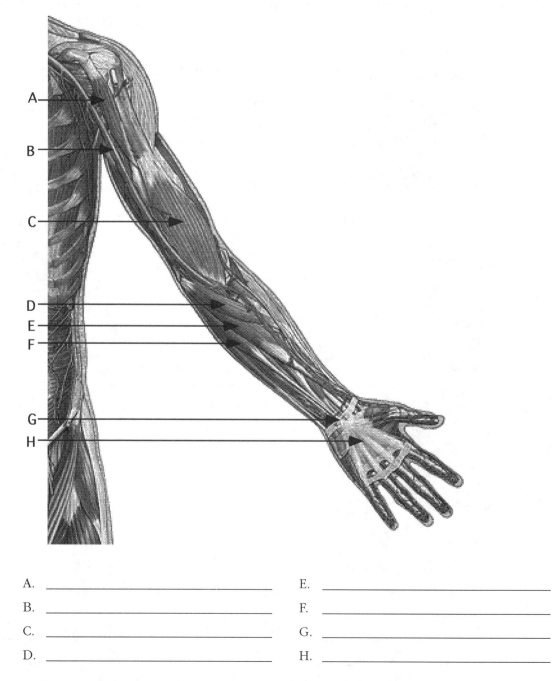

A. _____   E. _____

B. _____   F. _____

C. _____   G. _____

D. _____   H. _____

2. What does the name *coracobrachialis* imply about this muscle?

   a. PA (origin)? _____

   b. DA (insertion)? _____

3. The brachialis muscle is a powerful elbow flexor. What is the derivation of *brachialis*?

   _____

4. The first part of the name, *pronator*, in the muscle pronator teres describes its action on the forearm. What is its action?

   _____

Exercise 2.8 Deep Muscles of the Anterior Forearm

➤ *Select Anterior from the View button drop-down menu. Adjust the Layer Indicator to 131.*

1. Observe the left forearm and wrist. Label the muscles on the diagram below.

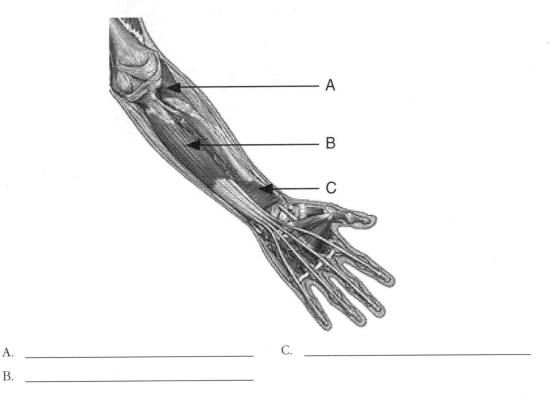

— A

— B

— C

A. _____   C. _____

B. _____

2. From its name, what can you imply about the action of flexor digitorum profundus?

   _____

3. Pronator quadratus is the prime mover for forearm pronation. What is the definition of a prime mover?

   _____

SUPERFICIAL MUSCLES OF THE ABDOMINAL REGION AND LOWER BODY

Exercise 2.9 Superficial Muscles of the Abdominal Region and Associated Structures

➤ *Using Dissectible Anatomy, adjust the image on your screen so that it matches the following diagram. Set the Layer Indicator to 20.*

1. Label the muscles and associated structures.

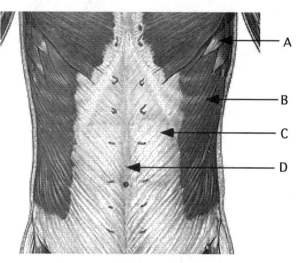

A. _____ C. _____

B. _____ D. _____

Beyond A I A

2. What is the derivation of *linea alba*?

_____

3. What is the function of the linea alba?

_____

4. What is an *aponeurosis*?

_____

Exercise 2.10 Deep Muscles of the Thorax and Abdomen

➤ *Adjust the Layer Indicator to 145.*

1. Label the muscles and associated structures on the diagram below.

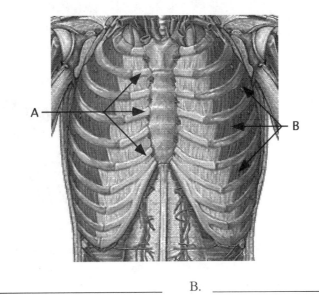

A. _____   B. _____

Beyond
A I A   2. What is the derivation of *intercostal?*

_____

3. Why is "intercostal" an appropriate name for these muscles?

_____

   (Note: These muscles elevate the rib cage during quiet inspiration.)

➤ *Adjust the Layer Indicator to 146.*

4. What are the muscles, lying deep to the external intercostals, that now fill the intercostal spaces in this view?

_____

Beyond
A I A   5. These muscles have the opposing action of the external intercostals. What is this action?

_____

> *Adjust the Layer Indicator to 169.*

6. Label the muscles and associated structures in the diagram below.

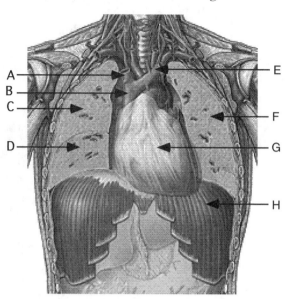

A. _____   E. _____

B. _____   F. _____

C. _____   G. _____

D. _____   H. _____

7. If you attempt to identify any of the abdominal organs over which the diaphragm lies, what term appears on the screen?

_____

8. To what anatomical class of membranes does your answer to "7" belong?

_____

9. The tendon at the top of the diaphragm is more sheetlike than cordlike. What is the name given to a broad, sheetlike tendon?

_____

10. What organs lie on the superior, lateral sides of the diaphragm?

_____

11. What organ lies on the central portion of the dome of the diaphragm?

_____

12. What is the action of the diaphragm?

_____

➢ *Adjust the Layer Indicator to 26. (In this view, the overlying external abdominal oblique has been removed.)*

13. Label the muscles and associated structures in the diagram below.

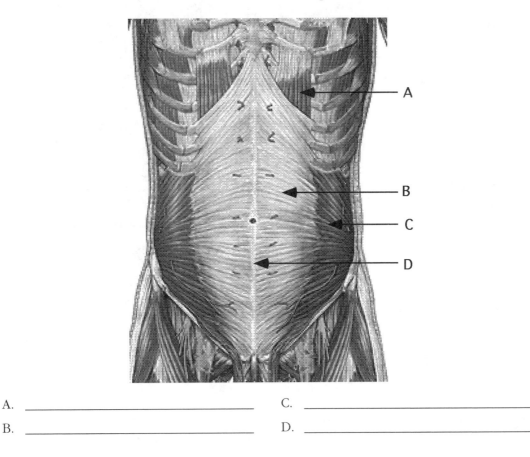

A. _____   C. _____

B. _____   D. _____

14. What name is given to the large, white central expansion that binds the fibers of the internal abdominal oblique muscle to the linea alba?

_____

15. What muscle, extending upward to attach to the costal cartilage of the fifth rib, lies just deep to the internal abdominal oblique muscles in this view?

_____

➢ *Adjust the Layer Indicator to 29.*

16. Label the muscles and associated structures in the diagram below.

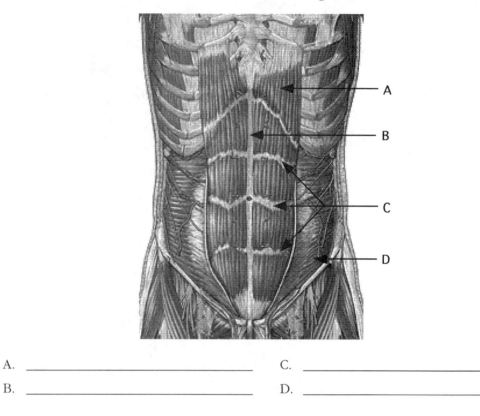

A. _____   C. _____

B. _____   D. _____

17. Describe the direction of the transversus abdominis muscle's fibers?

_____

18. What special midline structure "separates" the two vertical portions of the rectus abdominis muscle?

_____

19. What is the name of the four white, transverse bands running across the rectus abdominis muscle?

_____

(Note: A combined action of the muscles of the abdominal wall is compression of the abdominal viscera during defecation and micturition.)

➢ *Adjust the Layer Indicator to 266.*

20. Label the muscles and associated structures in the diagram below.

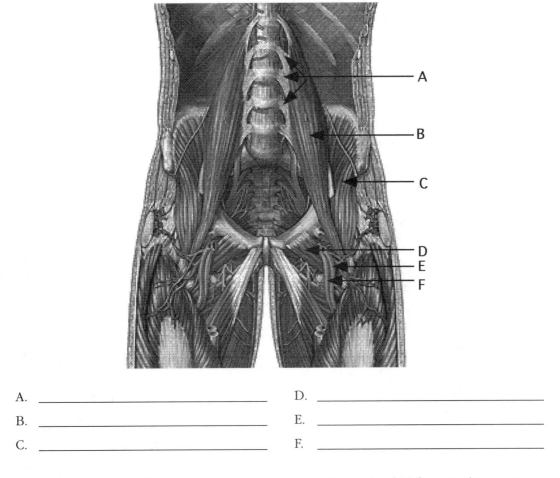

A. _____   D. _____

B. _____   E. _____

C. _____   F. _____

21. What muscle lies just lateral to the psoas major muscles in the iliac fossa in this view?

_____

22. What two major blood vessels cross the belly of the pectineus in this view?

_____

(Note: Both psoas major and iliacus strongly flex the hip and trunk and are assisted by the pectineus, which also adducts the hip.)

Exercise 2.11 Anterior Thigh Muscles and Associated Structures

➤ *Open Dissectible Anatomy and choose Male, Anterior. Click Open.*

➤ *Expand the image and adjust the Layer Indicator to 181. Zoom in on the image's right thigh.*

➤ *Select Open Content from the File menu, and click on Atlas Anatomy. Select Lower Limb from the Region menu, Muscular from the System menu, Anterior from the View menu, and Cadaver Photograph from the Type menu.*

➤ *Select the "Anterior Thigh" thumbnail icon. Click Open. Select Tile Vertically from the Window menu.*

1. Label the following images.

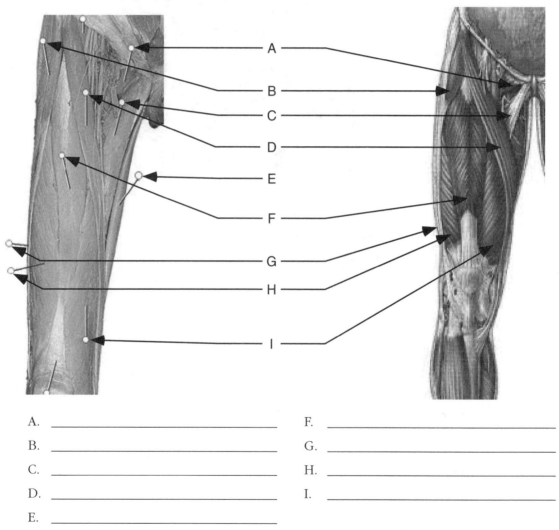

A. _____   F. _____

B. _____   G. _____

C. _____   H. _____

D. _____   I. _____

E. _____

2. What is the derivation of *sartorius*?

_____

3. What four muscles make up the quad muscles of the thigh?

_____

4. Which quad muscle is missing in the image on your screen? Why?

_____

5. For what two reasons is the quadriceps femoris muscle known by this name?

_____

6. To what bony structure does the patellar ligament attach?

_____

➤ *Click the Normal button on the Dissectible Anatomy window. Note the triangular region on the superior thigh where a vein (blue), artery (red), and nerve (yellow) can be seen. This region is known as the femoral triangle and is bordered by the inguinal ligament, sartorius muscle, and adductor longus muscle. The vein is the femoral vein.*

7. The artery is the _____

8. The nerve is the _____

9. What is the name of the muscle that forms the lateral boundary of the femoral triangle?

_____

10. What is the name of the muscle that forms the medial boundary of the femoral triangle?

_____

11. What forms the superior boundary of the femoral triangle?

_____

Exercise 2.12 Deep Anterior/Medial Thigh Muscles and Associated Structures

➤ *Using Dissectible Anatomy, Anterior view, adjust the Layer Indicator to 266.*

1. Adjust the image to match the diagram below, and then label the muscles and associated structures.

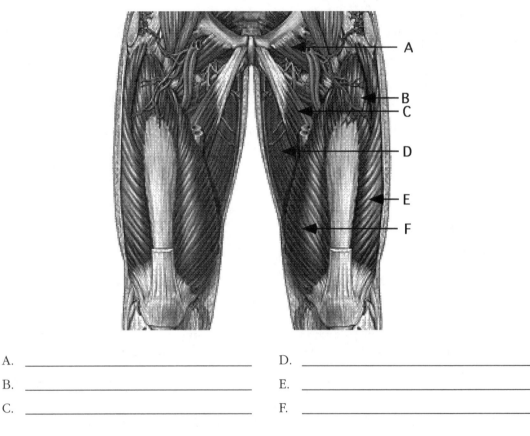

A. _____   D. _____

B. _____   E. _____

C. _____   F. _____

2. What muscle lies just inferiomedially to the pectineus in this view?

_____

3. What two superficial anterior thigh muscles are missing in this view?

_____

Beyond A I A 4. What is the main action of the medial thigh muscles?

_____

Exercise 2.13 Posterior Buttock, Thigh, and Leg Muscles with Associated Structures

➤ *Select Dissectible Anatomy, Posterior view, and adjust the Layer Indicator to 11. Zoom in on the image's right buttock and thigh.*

➤ *Select Open Content from the File Menu and choose Atlas Anatomy.*

> *Select Lower Limb from the Region menu, Muscular from the System menu, Posterior from the View menu, and Cadaver Photograph from the Type menu.*

> *Select the "Dissection of Post . . ." thumbnail icon. Click Open.*

> *Select Tile Vertically from the Window menu, and adjust the image to match the following illustration.*

1. Label the muscles and associated structures.

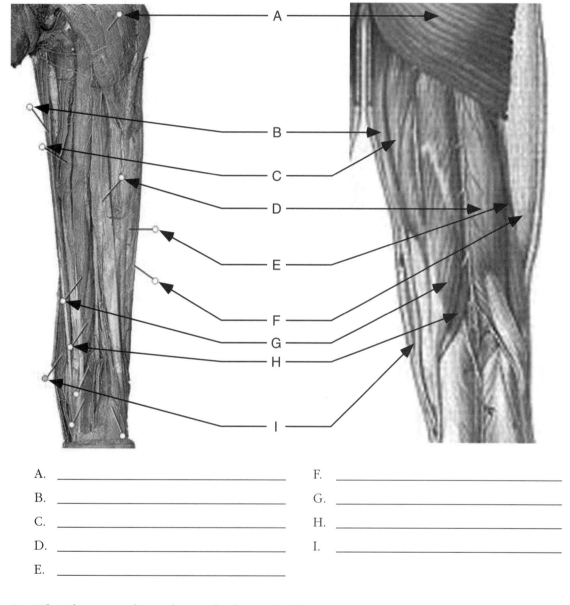

A. _____   F. _____

B. _____   G. _____

C. _____   H. _____

D. _____   I. _____

E. _____

2. What three muscles make up the hamstrings?

_____

Beyond A I A
3. What two facts can be inferred from the name *biceps femoris*?

_____

➤ *Close the Atlas Anatomy window, zoom out on the image, and adjust the image to match the diagram below.*

4. Use the Identify tool to label the diagram.

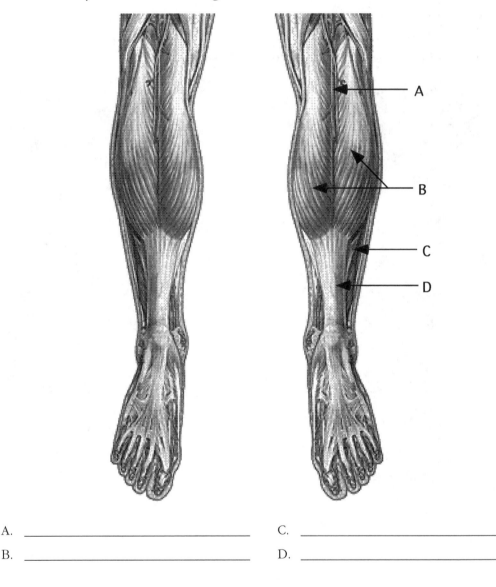

A. _____   C. _____

B. _____   D. _____

Beyond
A I A

5. What is the action of the gastrocnemius muscle?

_____

6. What is the distal attachment (insertion) of the gastrocnemius muscle?

_____

Exercise 2.14 Deep Muscles of the Buttock

➤ *Open Dissectible Anatomy, choose a Lateral view, and set the Layer Indicator to 93. Zoom in on the hip region.*

➤ *Select Open Content from the File menu and choose Atlas Anatomy.*

➤ *Select Lower Limb from the Region menu, Muscular from the System menu, Lateral from the View menu, and Cadaver Photograph from the Type menu.*

> ➤ *Select the "Dissection of Glut . . . " thumbnail icon. Click Open.*

> ➤ *Select Tile Vertically from the Window menu. Adjust the image to match the following illustration.*

1. Label the muscles and associated structures.

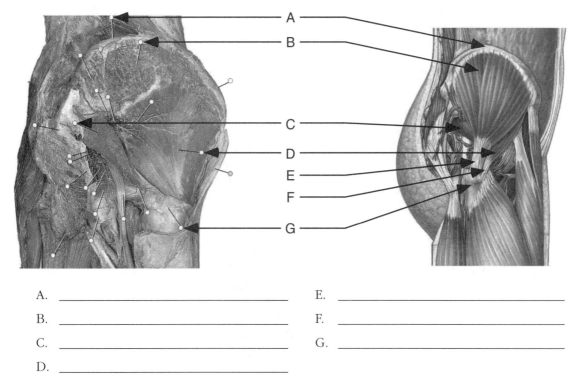

A. _____   E. _____

B. _____   F. _____

C. _____   G. _____

D. _____

> ➤ *Close the Atlas Anatomy window, expand the Dissectible Anatomy window, and choose a Posterior view. Adjust the Layer Indicator to 108.*

2. Label the muscles and associated structures in the following diagram.

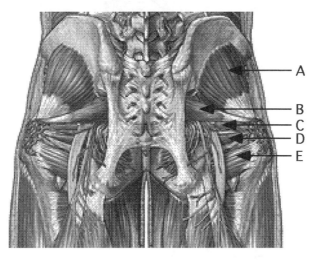

A. _____   D. _____

B. _____   E. _____

C. _____

3. What muscle lies just superficial to the gluteus minimus?

_____

4. The muscles lettered B, C, D, and E, identified in the previous diagram, are deep external rotators of the hip. On what femoral structure do they all insert?

_____

Exercise 2.15 Lateral Leg Muscles

➤ *Select Lateral from the View button drop-down menu. Center the image on the right lateral leg and foot. Adjust the Layer Indicator to 62.*

1. Label the diagram below.

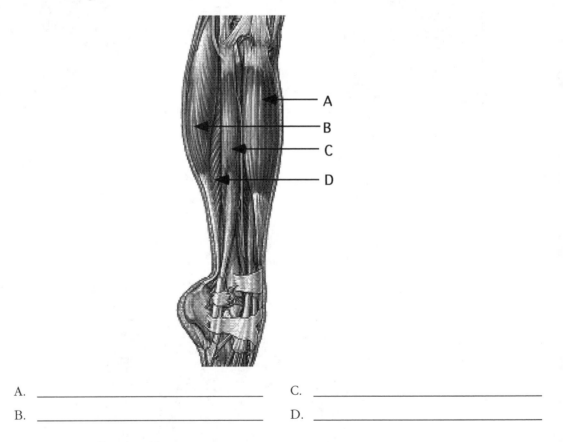

A. _____   C. _____

B. _____   D. _____

2. What is the name of the nerve running between the bellies of the peroneus longus and the tibialis anterior?

_____

Beyond
A I A

3. What is the action of the peroneus longus muscle?

_____

Exercise 2.16 Deep Lateral Leg Muscles

➢ *Select a Lateral view. Adjust the Layer Indicator to 88.*

1. Adjust the image to match the diagram below. Label the diagram.

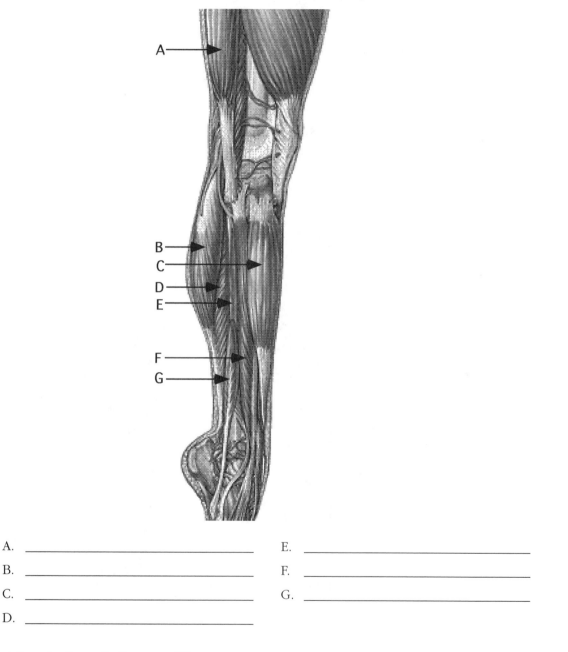

A. _____   E. _____

B. _____   F. _____

C. _____   G. _____

D. _____

➢ *Adjust the Layer Indicator to 87.*

2. What muscle is superficial to the peroneus brevis?

_____

(Note: Both the peroneus longus and the peroneus brevis are strong everters of the foot.)

**Beyond
A I A**

3. What is the opposite movement of eversion?

_____

4. In terms of the direction of the sole of the foot, describe

   a. inversion: _____

   b. eversion: _____

5. In the previous diagram, what ankle joint movement is depicted?

   _____

6. Which two muscles shown in the diagram are the prime movers for plantar flexion?

   _____

Exercise 2.17 Anterior Leg Muscles

➤ *Select an Anterior view. Adjust the Layer Indicator to 182.*

1. Adjust the image to match the diagram below. Label the diagram.

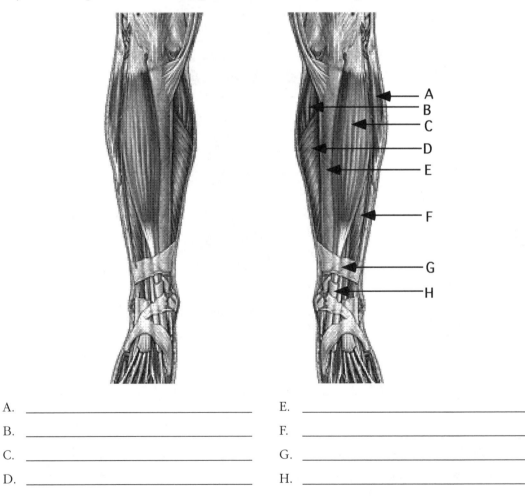

A. _____   E. _____

B. _____   F. _____

C. _____   G. _____

D. _____   H. _____

Beyond
A I A 2. What does *tibialis anterior* mean?

_____

3. What is the action of the tibialis anterior?

_____

4. What action can be inferred from the name *extensor digitorum longus?*

_____

5. What is the function of a tendon sheath?

_____

Exercise 2.18 Deep Anterior and Posterior Leg Muscles

➤ *Select an Anterior view. Adjust the Layer Indicator to 187.*

1. Adjust the image to match the following diagram. Label the diagram.

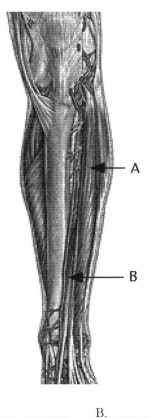

A. _____   B. _____

Beyond
A I A 2. The action of extensor digitorum longus is to extend all the joints of the lateral four toes. Follow the tendon of extensor hallucis longus onto the foot. What is the action of this muscle?

_____

➤ *Select a Posterior view. Adjust the Layer Indicator to 153.*

3. Label the following diagram.

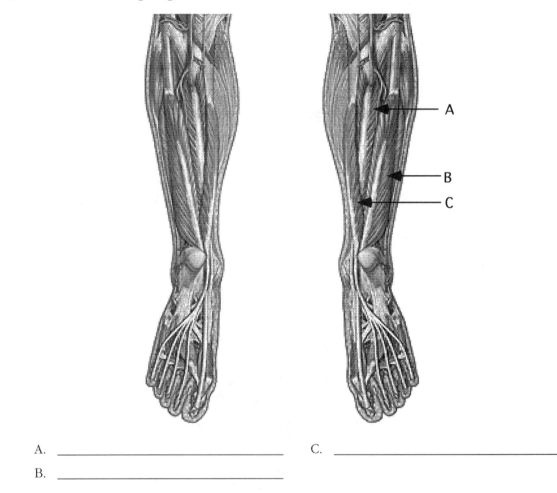

A. _____   C. _____

B. _____

4. Which of the three deep muscles in the diagram above is the most centrally located?

_____

5. Most medially located?

_____

6. Most laterally located?

_____

Beyond
A I A

7. What does the name *flexor digitorum longus* imply about its action?

_____

8. What does the term *tibialis posterior* imply about its specific location?

_____

(Note: The tibialis posterior both plantar flexes the ankle and inverts the foot.)

3

NERVOUS SYSTEM

STUDENT OBJECTIVES

OVERVIEW
- Review the anatomy of the brain and cranial nerves, spinal cord and spinal nerves, and autonomic nervous system.

BRAIN AND CRANIAL NERVES
- Describe the surface features of the brain, including gyri, sulci, and lobes.
- Identify major regions of the brain from a lateral and medial view.
- Locate and describe the 12 pairs of cranial nerves.

SPINAL CORD AND SPINAL NERVES
- Describe the anatomy of the spinal cord, especially in cross section, and be able to identify the anterior and posterior gray horns; the anterior, lateral, and posterior white columns; and the gray commissure.
- Describe the structure and function of the meningeal layers.
- Identify the components of a spinal nerve, including the ventral and dorsal root components.

- Determine the method of numbering and naming spinal nerves.
- Name and describe some important nerves and blood vessels associated with the cervical, brachial, and lumbosacral plexuses.

AUTONOMIC NERVOUS SYSTEM
- Describe the formation of the sympathetic trunk.
- Describe some major ganglia and associated nerves of the sympathetic chain.
- Distinguish between preganglionic and post-ganglionic neurons and the significance of these neuron types.
- Distinguish between gray and white rami communicans.
- Describe the formation of the pterygopalatine ganglion.
- Identify major branches of the pterygopalatine ganglion and their distribution.
- Relate the anatomical differences between the sympathetic and parasympathetic divisions of the autonomic nervous system.

BRAIN AND CRANIAL NERVES

Exercise 3.1 Cerebrum and Cerebellum

➤ *Open AIA by double-clicking Start Interactive Anatomy. Select Dissectible Anatomy and the Lateral view thumbnail icon. Click Open. Expand the window. Zoom in on the head.*

➤ *Set the Layer Indicator to 191.*

1. A gyrus is an elevated region on the cerebrum. What gyrus lies directly posterior to the precentral gyrus?

   _____

2. A sulcus is a grooved region on the cerebrum separating individual gyri. What is the name of the sulcus that separates the precentral and postcentral gyri?

   _____

3. Locate and label the structures/regions listed below on the following diagram:

   - Precentral gyrus
   - Postcentral gyrus
   - Cerebellum
   - Central sulcus

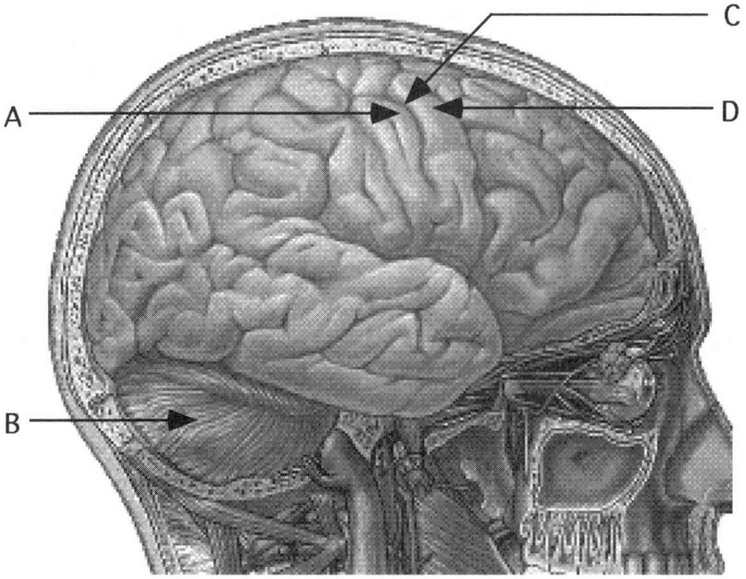

   A. _____   C. _____
   B. _____   D. _____

➤ *Select Medial from the View button drop-down menu. The Layer Indicator should be set to 99.*

4. Label the structures listed below on the following diagram. (Note: The entire right half of the cerebrum seen in this view is the right cerebral hemisphere.)

   - Frontal lobe
   - Occipital lobe
   - Parietal lobe
   - Cerebellum

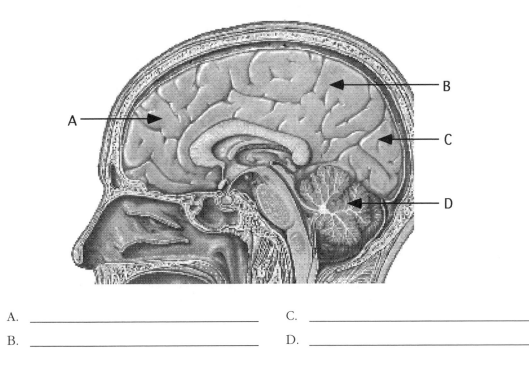

A. _____   C. _____

B. _____   D. _____

Exercise 3.2 Cerebrospinal Fluid (CSF)

1. Use the Identify tool to help you label the terms listed below on the following diagram:

- Corpus callosum
- Fornix
- Septum pellucidum

- Fourth ventricle
- Cerebral aqueduct
- Third ventricle

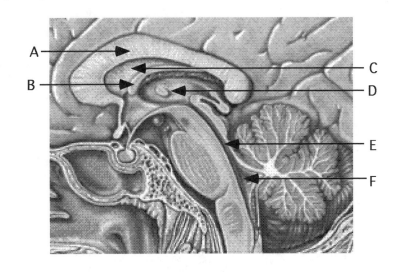

A. _____   D. _____

B. _____   E. _____

C. _____   F. _____

2. What is the name of the matter that is colored red and inferior to the fornix?

_____

3. What is the function of the corpus callosum?

_____

4. What is the function of the choroid plexus ?

_____

Exercise 3.3 Brain Structures

1. Use the Identify tool with the Medial view of the head to label the following terms on the diagram below. (Note: The terms in parentheses represent the actual labels in AIA and are listed to help you correlate these specific anatomical regions to their more general counterparts.)

- Medulla oblongata
  (Inferior olivary nucleus)

- Optic chiasma

- Thalamus
  (Interthalamic adhesion)

- Hypothalamus (Hypothalamic
  part of third ventricle)

- Pineal gland

- Mammillary body

- Cerebellum

- Superior colliculus

- Inferior colliculus

- Cerebrum
  (Precentral gyrus)

- Adenohypophysis

- Pons

- Neurohypophysis

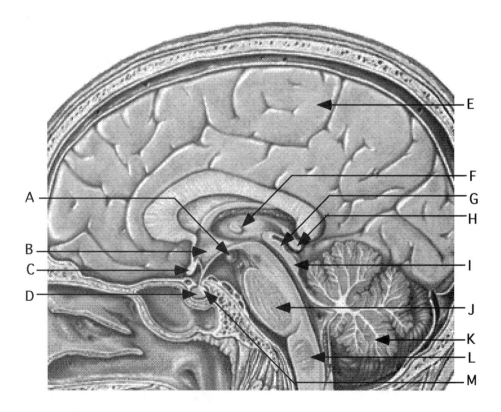

A. _____   H. _____

B. _____   I. _____

C. _____   J. _____

D. _____   K. _____

E. _____   L. _____

F. _____   M. _____

G. _____

Beyond A I A 2. Together, the superior and inferior colliculi make up what structure?

_____

3. What is the *arbor vitae*?

_____

4. What is the significance of the sella turcica in the sphenoid bone?

_____

5. What does *optic chiasma* mean, literally?

_____

6. The pituitary gland is subdivided into the anterior pituitary and posterior pituitary.

a. In AIA, the anterior pituitary is called what?

_____

b. In AIA, the posterior pituitary is called what?

_____

Beyond A I A 7. What is the function of the structure known as the infundibulum in the brain?

_____

8. What two hormones are made by the hypothalamus and stored in the posterior pituitary?

_____

9. What is the function of the cerebellum?

_____

10. What is the function of the medulla oblongata?

_____

Clinical Animation

To observe a clinical animation of the brain and some of its functions, do the following:

➢ *Select Open Content from the File Menu. Select Clinical Animations. From the drop-down menus which appear, make the following selections:*

Body Region: Head and Neck
Body System: Nervous
Medical Specialty: Neurology

➤ *Click on the "Brain Components" thumbnail icon. Click Open and expand the window by clicking the Maximize button.*

➤ *A "Brain Components" window will open and begin to play. Below the animation window a text window can be scrolled through.*

➤ *After the animation finishes, close the "Brain Components" window.*

Exercise 3.4 Cranial Nerves

➤ *Select Open Content from the File menu.*

➤ *Click Atlas Anatomy. From the Region menu, select Head of Neck. From the System menu, select Nervous. From the View menu, select Inferior.*

➤ *Select "Cerebral Arterial Circle (Inf)" from the two thumbnail icons. Click Open.*

➤ *The image that appears shows the base of the brain and its associated cranial nerves and blood vessels. A special anastomosis of the four arteries that supply blood to the brain, known as the circle of Willis (cerebral arterial circle), is also visible. If you select the head of a pin, it turns green and an identification tag appears next to it. The structure's name also appears in the Structure List at the top of the window.*

➤ *From the Systems drop-down menu icon, select Nervous. Now only the pins associated with the nervous system appear on the image.*

1. Label the lettered structures in the following diagram. (Note: Letters P and Q have been added to the diagram as they were not identified in AIA. You may need to use reference materials to label these pins.)

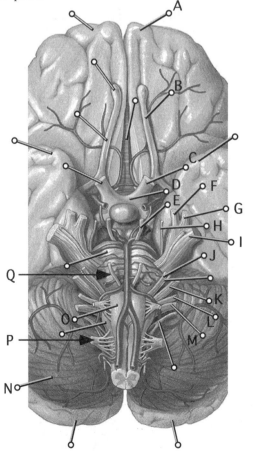

A. _____

B. _____

C. _____

D. _____

E. _____

F. _____

G. _____

H. _____

I. _____

J. _____

K. _____

L. _____

M. _____

N. _____

O. _____

P. _____

Q. _____

2. Fill in the following chart. You may need to refer to your text for the function of each nerve.

Cranial Nerve Name	Cranial Nerve Number	Function (Sensory, Motor, Both)
	I	
	II	
	III	
	IV	
	V	
	VI	
	VII	
	VIII	
	IX	
	X	
	XI	
	XII	

Exercise 3.5 Cranial Nerves: AIA 3D

➢ *Close the Atlas Anatomy and Dissectible Anatomy windows.*

➢ *Select Open Content from the File menu. Click 3D Anatomy and select Brain. Click Open.*

➢ *Maximize the view on your screen and increase the magnification.*

1. Using the Structure List for Brain at the top of your screen, scroll down the list, and label the following structures on the diagram below. (Note: The structures are listed in alphabetical order.)

- Abducent (Abducens) nerve {CN VI}
- Accessory nerve {CN XI}
- Facial nerve {CN VII}
- Glossopharyngeal nerve {CN IX}
- Hypoglossal nerve {CN XII}
- Medulla oblongata
- Oculomotor nerve {CN III}
- Olfactory nerve {CN I}

- Optic chiasma
- Optic nerve {CN II}
- Pons
- Trigeminal nerve {CN V}
- Trochlear nerve {CN IV}
- Vagus nerve {CN X}
- Vestibulocochlear nerve {CN VIII}

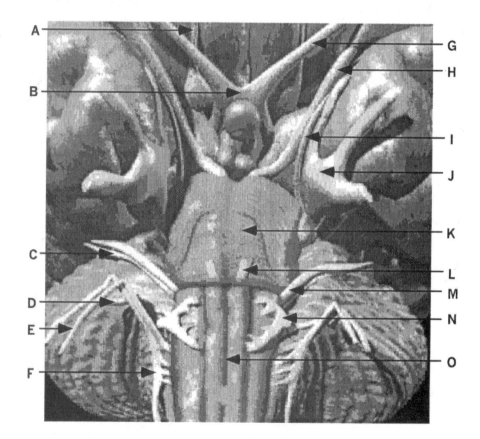

A. _____   I. _____

B. _____   J. _____

C. _____   K. _____

D. _____   L. _____

E. _____   M. _____

F. _____   N. _____

G. _____   O. _____

H. _____

SPINAL CORD AND SPINAL NERVES

Spinal Cord
Exercise 3.6 Meninges

➤ *Open Dissectible Anatomy.*

➤ *Select the Posterior thumbnail icon from the View window. Click Open.*

➤ *Expand the window and set the Layer Indicator at 178.*

➤ *If the image that appears on your screen does not display the entire head, neck, and back region, zoom out, using the Zoom button in the toolbar.*

➤ *The figure shows the entire spinal cord extending from the base of the skull down to its most inferior extent, the filum terminale. (Note: The vertebral laminae and spinous processes have been extracted, revealing the spinal cord and its outermost covering, the dura mater.)*

➤ *Adjust the Layer Indicator to 179, and click the spinal cord.*

1. What name appears at the top of the window?

_____

Beyond AIA
2. What is the literal meaning of the term *dura mater*, and what does it tell you about this layer of the meninges?

_____

3. Of what specific type of tissue does the dura consist?

_____

4. The arachnoid is the intermediate layer of the meninges. What does its name mean, literally?

_____

5. The deepest meningeal layer is not identified by AIA but is known as the pia mater. What does *pia mater* mean, literally?

_____

6. Of what specific type of tissue does the pia mater consist?

_____

Exercise 3.7 Meningeal Cavities

Beyond AIA
1. The space above the dura mater is the _____
2. What does this space contain? _____
3. The space below the dura mater is the _____
4. What does this space contain? _____
5. The space below the arachnoid is the _____
6. What does this space contain? _____

7. Describe cerebrospinal fluid in terms of the following:

    a. Its cellular components: _____

    b. Its chemical composition: _____

    c. Its function: _____

Exercise 3.8 Spinal Cord

➤ *Adjust the Layer Indicator to 180, and then zoom in on the lower region of the rib cage.*

➤ *Use the Identify tool to identify various dorsal root ganglia (seen exiting the spinal cord). Locate the spinal (dorsal root) ganglion of the spinal nerve.*

➤ *Once you have located the dorsal root ganglion of T10, highlight this structure by clicking the Highlight button in the toolbar.*

1. Use the Identify tool to label the diagram below with the following terms:

- Ventral nerve root of T10
- Spinal (dorsal root) ganglion of T10
- Dorsal ramus of T10
- Gray ramus communicans of T10
- White ramus communicans of T10

- Dorsal nerve root of T10
- Ventral ramus of T10
- Pedicle of T10 vertebra
- Pedicle of T11 vertebra

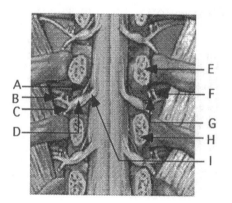

A. _____   F. _____

B. _____   G. _____

C. _____   H. _____

D. _____   I. _____

E. _____

2. In general, the dorsal rami will innervate what part of the body?

_____

3. In general, the ventral rami will innervate what part of the body?

_____

4. What type of nerve (sensory, motor, mixed) is found within the ventral root?

_____

5. What type of nerve (sensory, motor, mixed) is found within the dorsal root?

_____

6. What type of nerve (sensory, motor, mixed) is the spinal nerve?

_____

7. In what structure are cell bodies of sensory neurons located?

_____

8. Within what region of the spinal cord's gray matter are motor neuron cell bodies located?

_____

9. To what structure do the gray and white communicating rami lead?

_____

10. Adjust the image on your screen to the most inferior extent of the spinal cord so that it matches the diagram that follows. (The Layer Indicator should remain at 180.) Label the diagram below.

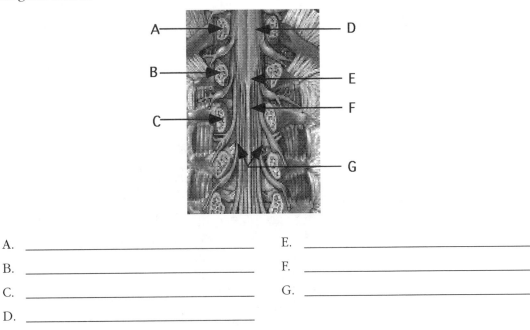

A. _____   E. _____

B. _____   F. _____

C. _____   G. _____

D. _____

11. Between which two vertebrae does the spinal cord end?

_____

12. By what name is the inferior, tapered end of the spinal cord known?

_____

13. By what name is the long, slender strand extending from the inferior end of the conus medullaris and attaching to the coccyx known?

_____

14. What does *cauda equina* mean, literally?

_____

15. Why do you suppose that anatomists chose the term *cauda equina* to identify this structure?

_____

16. In respect to a vertebra, what is a *lamina*?

_____

Exercise 3.9 Spinal Cord and Meninges

➤ *Select Open Content from the File menu. Click the Atlas Anatomy button. Select the following from the indicated drop-down menus:*

Region: Body Wall and Back   View: Non-standard
System: Nervous   Type: Illustration

➤ *Select "Spinal Cord Vess . . . " thumbnail icon.*

➤ *Click Open. A new window, Spinal Cord Vessels and Meninges, appears. Expand the window.*

➤ *Zoom out, if necessary and select Nervous from the system drop-down menu.*

1. Label the lettered structures on the following diagram.

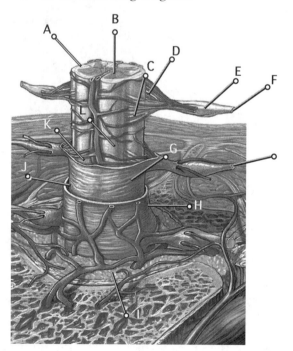

A. _____	G. _____
B. _____	H. _____
C. _____	I. _____
D. _____	J. _____
E. _____	K. _____
F. _____	L. _____

Exercise 3.10 Superior View of the Spinal Cord Within a Vertebra

➤ *Select Open Content from the File menu. Atlas Anatomy should be selected. Select the following from the indicated drop-down menus:*

Region: Body Wall and Back   *View: Superior*
System: Nervous   *Type: Illustration*

➤ *Select the "T12 Vertebra (Sup)" thumbnail icon. Click Open. A new window, T12 Vertebra (Sup), appears.*

➤ *Expand the window and zoom out, if necessary, using the Zoom button in the Tool palette.*

1. Select Show All Pins from the Tool palette, and then label the lettered structures on the following diagram.

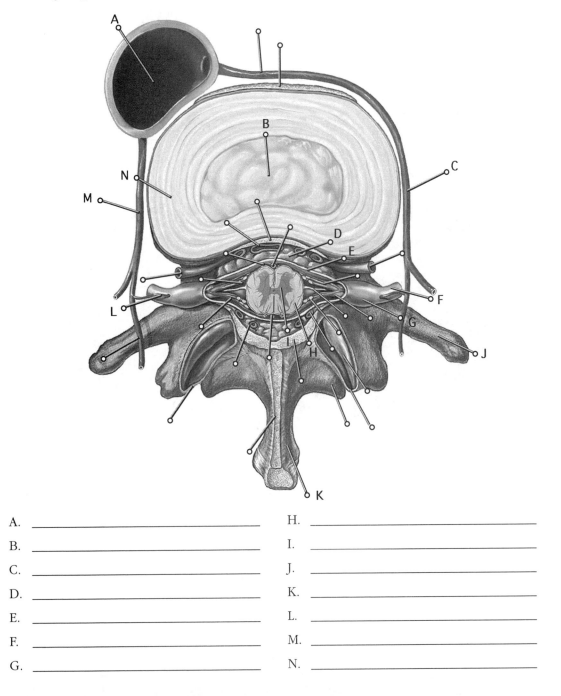

A. _____	H. _____
B. _____	I. _____
C. _____	J. _____
D. _____	K. _____
E. _____	L. _____
F. _____	M. _____
G. _____	N. _____

2. How are the terms *annulus fibrosus* and *nucleus pulposus* related to the structure of an intervertebral disc?

_____

3. What does the term *intercostal,* as in *intercostal arteries,* describe with respect to anatomical distribution?

_____

➤ *Zoom in and center the image so that it matches that of the diagram that follows.*

4. Label the lettered structures on the diagram. (Note: Letters A, B, C, D, and L have been added to the diagram as they were not identified in this image. You may need to use reference materials to label them.)

- Dura mater
- Epidural fat
- Subarachnoid space
- Spinal nerve
- Gray matter
- White matter
- Anterior gray horn

- Posterior gray horn
- Spinal (dorsal root) ganglion of spinal nerve
- Dorsal nerve root of spinal nerve
- Ventral root of spinal nerve
- Gray commissure
- Central canal

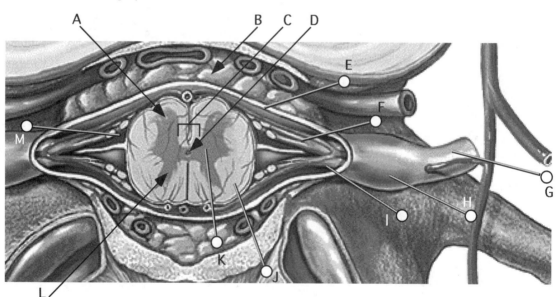

A. _____	H. _____
B. _____	I. _____
C. _____	J. _____
D. _____	K. _____
E. _____	L. _____
F. _____	M. _____
G. _____	

Spinal Nerves

Exercise 3.11 Terminal Branches of the Brachial Plexus

➤ *The brachial plexus is formed by the union of several ventral rami, typically C5 through T1, and supplies the muscles of the upper limb. Close Atlas Anatomy. Open Dissectible Anatomy.*

➤ *Select the Anterior view thumbnail icon. Click Open and expand window.*

➤ *Using the Zoom button, zoom in on the image.*

➤ *Using the Navigator icon, center the right upper arm and shoulder region in the window, and set the Layer Indicator to 82.*

1. What is the name of the muscle that is attached to the coracoid process of the scapula in this view?

   _____

2. What is the name of the nerve that passes through this muscle?

   _____

3. Follow this nerve as it travels distally the length of the arm. What muscle lies deep to it as this nerve enters the forearm?

   _____

4. Label the following diagram of the upper arm and shoulder region with the terms listed below:

   - Musculocutaneous nerve
   - Brachial artery
   - Basilic vein
   - Coracoid process of scapula
   - Coracobrachialis muscle
   - Median nerve
   - Ulnar nerve

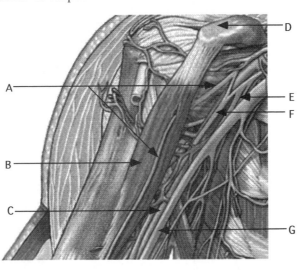

A. _____   E. _____

B. _____   F. _____

C. _____   G. _____

D. _____

5. Adjust the image on your screen to match that of the diagram below (distal portion of the right upper arm) and then label the diagram with the following terms:

- Musculocutaneous nerve
- Ulnar nerve
- Brachial artery
- Median nerve
- Basilic vein
- Brachialis muscle

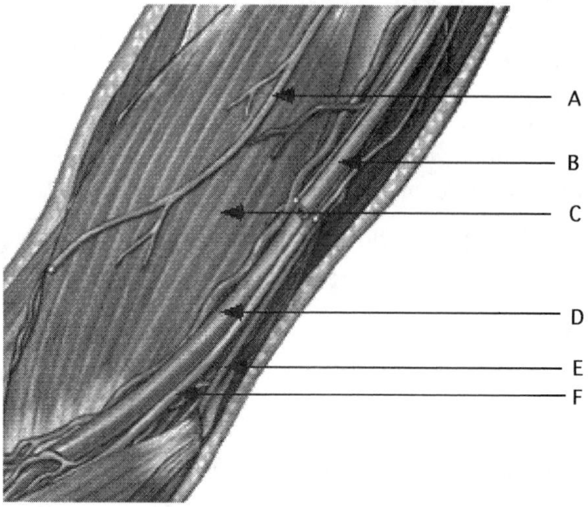

A. _____   D. _____

B. _____   E. _____

C. _____   F. _____

6. Which nerve is the most lateral in the diagram above?

_____

7. Which nerve is immediately medial to the musculocutaneous?

_____

8. Which nerve is the most medial in the diagram?

_____

9. What artery is parallel to the median nerve in this view?

_____

**Beyond
A I A** 10. What is the clinical significance of the brachial artery?

_____

11. What is a nerve plexus?

_____

12. What is the nerve plexus that supplies the muscles of the upper extremity?

_____

Exercise 3.12 Phrenic Nerve

➤ *Select Find from the Tools menu. Type "Phrenic nerve" and click Find.*

➤ *Select "Phrenic nerve" from the Structure list.*

➤ *Click the Show Result In button. Select Dissectible Anatomy and the Male Anterior thumbnail icon. Click Open. The Layer Indicator should be set to 168.*

**Beyond
A I A** 1. From what plexus does the phrenic nerve arise?

_____

2. What muscle does the phrenic nerve innervate? Knowing this, what effect would a lesion of this nerve have on this muscle's action? Explain.

_____

Exercise 3.13   Femoral Nerve

➤ *Using the Navigator icon, center the image over the upper right thigh region, and adjust the Layer Indicator to 181.*

➤ *Click the Normal button to return to full color.*

1. Adjust the image on your screen so that it matches the following diagram. Label the diagram with the following terms:

    - Femoral nerve
    - Femoral artery
    - Femoral vein

    - Inguinal ligament
    - Adductor longus muscle
    - Sartorius muscle

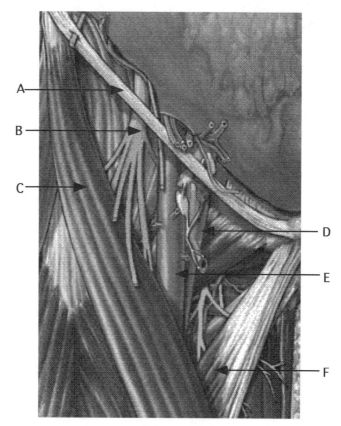

A. _____   D. _____

B. _____   E. _____

C. _____   F. _____

2. Two large blood vessels exit the abdominal cavity medially to the femoral nerve.

    a. The red vessel is _____

    b. The blue vessel is _____

Beyond
A I A

3. What is the name of the special triangular region bounded by the sartorius muscle, the inguinal ligament, and the adductor longus muscle?

_____

Exercise 3.14 Sciatic Nerve

➤ *Select Posterior from the View button drop-down menu.*

➤ *Set the Layer Indicator to 120, and zoom out.*

Beyond A I A

1. The sciatic nerve, the thickest nerve in the body, along with its subdivisions, the tibial and common peroneal nerves, supplies the posterior thigh and leg muscles. What collective name is given to the posterior thigh muscles that the sciatic nerve supplies?

_____

AUTONOMIC NERVOUS SYSTEM

Sympathetic Division

Exercise 3.15 Sympathetic Trunk

➤ *Select Anterior from the View button drop-down menu. Zoom in on the image.*

➤ *Set the Layer Indicator at 264, and using the Navigator tool, adjust the image to match the following illustration.*

1. Use the Identify tool to label the following diagram.

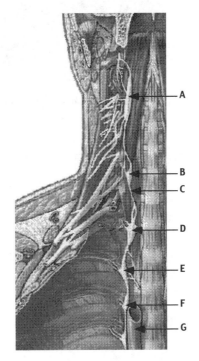

A. _____   E. _____

B. _____   F. _____

C. _____   G. _____

D. _____

2. What is the name of the large blue vessel running just medial to the right sympathetic chain ganglia?

_____

3. Why is the sympathetic chain also referred to as the paravertebral chain?

_____

4. What does the word *azygous* mean?

_____

(Note: The azygous system of veins on each side of the vertebral column drains the back and thoracic and abdominal walls. The azygous vein connects the superior and inferior vena cava either directly or by way of the hemizygous and accessory hemizygous veins. The azygous system is important because it offers an alternate route of venous drainage from the thoracic, abdominal, and back regions if the inferior vena cava is blocked.)

5. Use the Identify tool to label the diagram below.

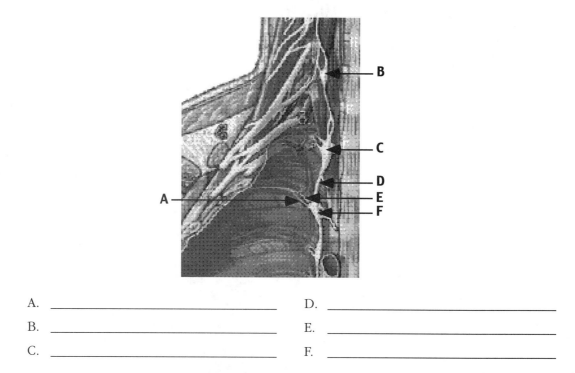

A. _____   D. _____

B. _____   E. _____

C. _____   F. _____

Exercise 3.16 Greater Splanchnic, Lesser Splanchnic, and Lumbar Splanchnic Nerves

➤ *Open Dissectible Anatomy and select Male and the Lateral view thumbnail icons.*

➤ *Click Open. Expand the window and adjust the Layer Indicator to 225. Zoom in on the image.*

➤ *Select Open Content from the File Menu. Click on the Anterior view thumbnail icon. Click Open.*

➤ *Adjust the Layer Indicator to 264 and use the Zoom tool to zoom in on the image.*

➤ *Adjust the image so that the fifth thoracic sympathetic ganglion is visible at the top of the window and matches that of the anterior view of the sympathetic chain in the following left diagram.*

➤ *Select Tile Vertically from the Window menu and adjust the right-hand image to match the figure below.*

1. Label the following diagrams. Identify each letter's structure and place its name in the appropriate space below the diagrams. (Note: The letters represent the same structure on both diagrams.)

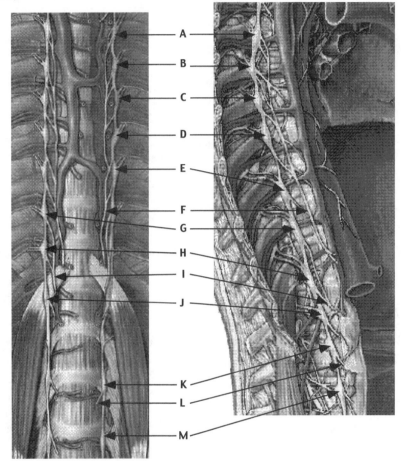

Anterior view of sympathetic chain   **Lateral view of sympathetic chain**

A. _____   H. _____

B. _____   I. _____

C. _____   J. _____

D. _____   K. _____

E. _____   L. _____

F. _____   M. _____

G. _____

Beyond
A I A
2. With respect to preganglionic fibers, describe the formation of the greater splanchnic nerve.

_____

3. With respect to preganglionic fibers, describe the formation of the lesser splanchnic nerve.

_____

4. With respect to preganglionic fibers, describe the formation of the lumbar splanchnic nerve.

_____

5. Does the white rami communicans contain fibers (axons) of a preganglionic or postganglionic neuron? How do you know?

_____

6. Does the gray rami communicans contain fibers (axons) of a preganglionic or postganglionic neurons? How do you know?

_____

7. Does the white rami communicans contain myelinated or unmyelinated axons of neurons?

_____

8. Does the gray rami communicans contain myelinated or unmyelinated axons of neurons?

_____

9. At what spinal cord locations do you find white rami communicans?

_____

10. White rami communicans connect what two structures?

_____

11. Gray rami communicans connect what two structures?

_____

Exercise 3.17 Celiac Ganglion, Superior Mesenteric Ganglion, and Inferior Mesenteric Ganglion

➤ *Close the Lateral Male window. Set the Layer Indicator to 239 on the Anterior Male window.*

1. Adjust the image on your screen to match the diagram that follows, and label the diagram.

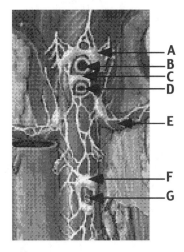

A. _____   E. _____

B. _____   F. _____

C. _____   G. _____

D. _____

2. What is the name of the red structure beneath the ganglia?

   _____

Beyond
A I A

3. In terms of the synapsing fibers, what is the celiac ganglion?

   _____

4. In terms of the synapsing fibers, what is the superior mesenteric ganglion?

   _____

5. In terms of synapsing fibers, what is the inferior mesenteric ganglion?

   _____

Parasympathetic Division
Exercise 3.18 Pterygopalatine Ganglion

➤ *Select Open Content from the File menu.*

➤ *Click Atlas Anatomy and click on Details at the top of the window. Select the following from the indicated drop-down menus:*

Region: Head and Neck
System: Nervous
View: Lateral
Type: Illustration

➤ *Click on Deails button at the top of the screen and select "Pterygopalatine Ganglion 1" at the bottom of the list. Click Open and expand the window.*

➤ *The image that appears displays the spinal cord and most of the brainstem (medulla oblongata and pons) with cranial nerves VII (facial) and V (trigeminal) emerging from the brainstem to supply, in part, the lacrimal gland, the nasal mucosa, and the palate.*

➤ *Select Nervous from the Systems drop-down menu.*

1. Label the following structures.

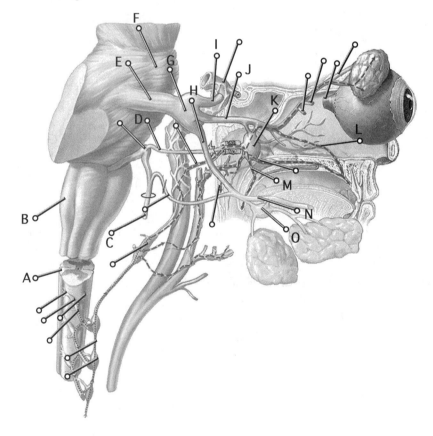

A. _____   I. _____

B. _____   J. _____

C. _____   K. _____

D. _____   L. _____

E. _____   M. _____

F. _____   N. _____

G. _____   O. _____

H. _____

2. Which division of the autonomic nervous system (sympathetic or parasympathetic) is depicted in the major portion of the diagram, including letters labeled C through H; how do you know?

   _____

3. Which major division of the trigeminal nerve receives postganglionic fibers emerging from the pterygopalatine ganglion?

   _____

4. Preganglionic fibers contributing to the formation of the pterygopalatine ganglion leave the skull by way of the greater petrosal nerve. The greater petrosal nerve is a branch of what cranial nerve?

   _____

Beyond AIA
5. What foramen does cranial nerve VII (facial nerve) pass through when it emerges from the skull?

   _____

6. Between what two temporal bone structures is the stylomastoid foramen located?

   _____

Beyond AIA
7. What parasympathetic ganglion sends postganglionic fibers to the sublingual and submandibular salivary glands?

   _____

8. What effect does parasympathetic innervation have on secretion of the lacrimal gland?

   _____

9. For what reason is the sympathetic division of the autonomic nervous system known as the thoracolumbar outflow?

   _____

10. What neurotransmitter is released at the preganglionic synapses of the sympathetic division?

   _____

11. What neurotransmitter is released at the postganglionic synapses of the sympathetic division?

   _____

12. From what specific brainstem region does the trigeminal nerve emerge?

_____

13. From what specific brainstem region does the facial nerve emerge?

_____

14. What neurotransmitter is released at the parasympathetic preganglionic synapse?

_____

15. What neurotransmitter is released at the parasympathetic postganglionic synapse?

_____

16. Describe the craniosacral outflow of the autonomic nervous system.

_____

➤ *Open Dissectible Anatomy, select Male Anterior, and adjust the Layer Indicator to 177.*

➤ *Zoom in on the image.*

17. Center on the upper chest and label the diagram below.

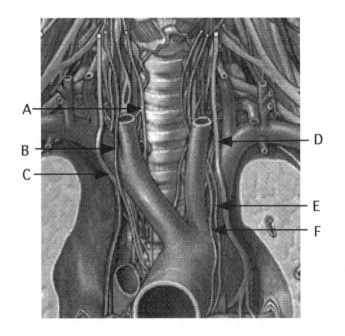

A. _____   D. _____

B. _____   E. _____

C. _____   F. _____

THE SPECIAL SENSES

STUDENT OBJECTIVES

OVERVIEW

- Review the anatomy of the associated structures related to vision, hearing, smell, and taste.

VISION

- Identify the extrinsic muscles of the eye.
- Describe the action of the extrinsic muscles of the eye.
- Identify the associated cranial nerve innervation of the extrinsic muscles of the eye.
- Identify the fibrous tunic of the eyeball, including the cornea and sclera.
- Identify the vascular tunic of the eyeball, including the choroid, ciliary body, iris, and pupil.
- Identify the inner layer of the eyeball, the retina, and its associated central fovea, optic disk, and optic nerve.
- Identify the structure of the eyeball's interior, including the lens, anterior chamber, and posterior chamber.

HEARING

- Identify the structure of the external ear.
- Identify the bones of the middle ear.
- Identify the structures of the inner ear.
- Identify the associated structures of the cochlea.

SMELL

- Describe the relationship of the olfactory nerve, olfactory bulb, and olfactory tract with the cribriform plate of the ethmoid bone.
- Identify the super, middle, and inferior nasal conchae.

TASTE

- Identify the various papillae of the tongue.

VISION

Exercise 4.1 Muscles of the Eye

➤ *Open AIA by clicking the Start Interactive Anatomy. Choose Atlas Anatomy.*

➤ *Select Head and Neck from the Body Region drop-down menu and Lateral from the View Orientation drop-down menu. Select the "Extrinsic Eye Muscles (Lat)" thumbnail icon. Click Open.*

➤ *Select Open Content from the File menu. Choose Atlas Anatomy. Make sure that Head and Neck and Lateral are still selected. Select the "Dissection of Orbit (Lat)" thumbnail icon. Click Open.*

➤ *Select Tile Vertically from the Window menu.*

1. Label the lettered pins on the diagram that follows.

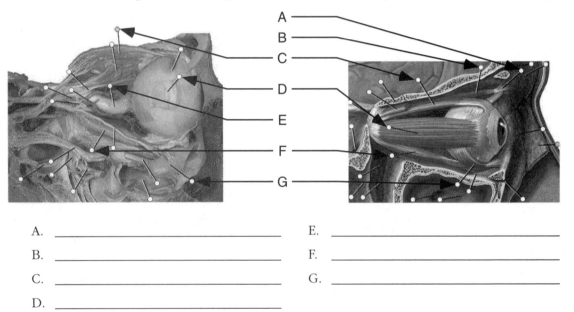

A. _____   E. _____

B. _____   F. _____

C. _____   G. _____

D. _____

2. What muscle passes through the trochlea?

_____

➤ *Close the Atlas Anatomy windows.*

➤ *Select Open Content from the File menu and choose 3D Anatomy. Select 3D Eye. Click Open.*

➤ *To determine the correct orientation of the muscles of the eye, it is helpful to know where the lacrimal gland is located. Choose "Lacrimal gland" from the Structure List by clicking the Structure List Bar at the top of the screen. Scroll down the list and select "Lateral rectus muscle." Compare the position of the lacrimal gland and the lateral rectus muscle. Select "Eyeball" to return to an anterior view.*

3. What is the anatomical position of the lacrimal gland with respect to the eye?

_____

➤ *To observe the actual contractions of the superior rectus muscle, select "Superior rectus muscle—Action." Note the resulting movement of the eye. This can be repeated for each of the muscles of the eye.*

4. Describe the movement of the eye, in directional terms, for each of the following:

   a. Lateral rectus _____

   b. Medial rectus _____

   c. Superior rectus _____

   d. Inferior rectus _____

   e. Superior oblique _____

   f. Inferior oblique _____

Exercise 4.2 Nerves of the Eye

Beyond
A I A

➤ *Choose "Trochlear nerve {CN IV}" from the 3D Eye Structure List.*

1. What muscle is innervated by this nerve?

_____

2. What is the relationship of this muscle and the name of the nerve?

_____

➤ *Choose "Abducent (Abducens) nerve {CN VI}" from the Structure List.*

3. What muscle is innervated by this nerve?

_____

➤ *Choose "Inferior division of oculomotor nerve {CN III}" from the Structure List.*

4. What muscle is innervated by this nerve?

_____

➤ *Choose "Superior division of oculomotor nerve {CN III}" from the Structure List.*

5. What muscle is innervated by this nerve?

_____

➤ *Choose "Optic nerve {CN II}" from the Structure List.*

6. What is the function of the optic nerve?

_____

➤ *Choose "Lacrimal nerve" from the Structure List.*

7. What division of the autonomic nervous system does this nerve belong to?

_____

Exercise 4.3 Structures of the Eye

➤ *Use the 3D Eye Structure List to label the following figures.*

1. Label the following structures in the diagram below:

   - Cornea
   - Pupil
   - Conjunctiva

   - Lacrimal gland
   - Sclera
   - Iris

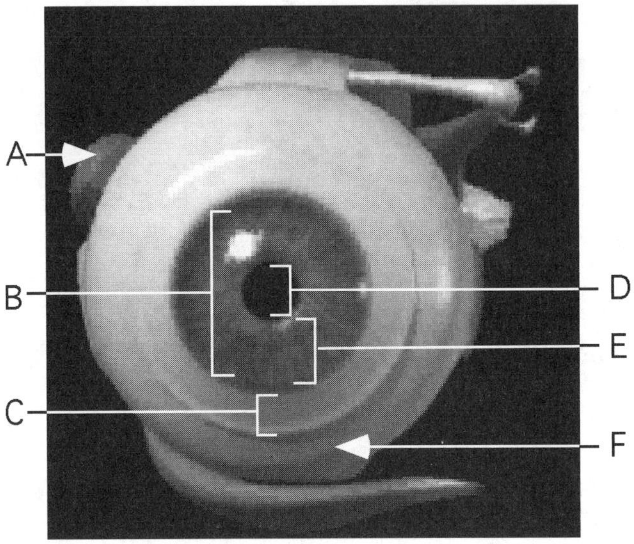

A. _____   D. _____

B. _____   E. _____

C. _____   F. _____

2. Use the 3D Eye Structure List to find and label the structures in the following diagram:

   - Anterior chamber of the eye
   - Posterior chamber of the eye
   - Central fovea of retina
   - Optic nerve {CN II}
   - Sclera (cut)

   - Ciliary body
   - Iris (cut)
   - Cornea (cut)
   - Lens
   - Ciliary process

A. _____   F. _____

B. _____   G. _____

C. _____   H. _____

D. _____   I. _____

E. _____   J. _____

Beyond
A I A 3. What is the significance of the central fovea?

_____

4. What is the significance of the macula?

_____

➤ *Using the 3D Eye Structure List, examine the position of the "Sclera (cut)," "Choroidea," and "Retina" with respect to each other. You may want to use the Zoom feature for this exercise.*

5. List the three structures from innermost to outermost.

a. Inner layer _____

b. Middle layer _____

c. Outer layer _____

Beyond
A I A 6. What is the function of the retina?

_____

7. What is the function of the choroidea?

_____

➤ *Close the 3D window.*

HEARING

Exercise 4.4 External, Middle, and Inner Ear

➤ *Select Open Content from the File menu. Choose Atlas Anatomy.*

➤ *Select Head and Neck from the Body Region drop-down menu and Superior from the View Orientation drop-down menu. Select the "External, Middle, and Inner Ear" thumbnail icon. Click Open.*

➤ *Expand the window to view the entire figure.*

1. Label the lettered pins on the diagram that follows.

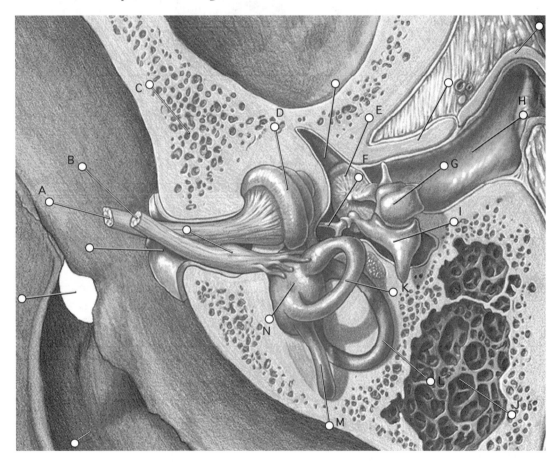

A. _____	H. _____
B. _____	I. _____
C. _____	J. _____
D. _____	K. _____
E. _____	L. _____
F. _____	M. _____
G. _____	N. _____

➤ *Close the Atlas Anatomy window.*

Exercise 4.5 Ossicles of the Middle Ear

➤ *Select Open Content from the File menu. Choose 3D Anatomy. Select 3D Ear and click Open. Expand the window.*

1. Label the following structures in the diagram below:

   • Malleus

   • Stapes

   • Incus

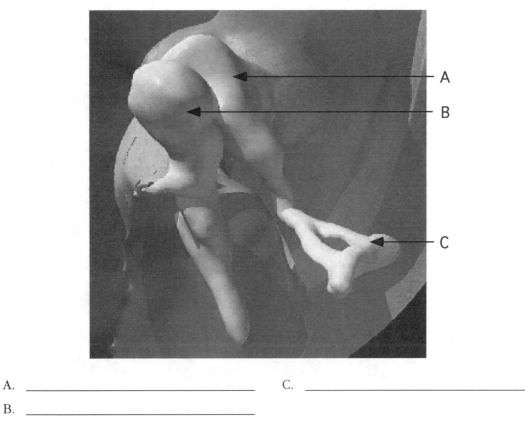

A. _____   C. _____

B. _____

Beyond
A I A

2. What is the common name of each of the following?

   a. Malleus _____

   b. Incus _____

   c. Stapes _____

Exercise 4.6 Inner Ear

1. Use the 3D Ear Structure List to label each of the following structures in the diagram below:

   - Lateral semicircular canal
   - Anterior semicircular canal
   - Posterior semicircular canal
   - Vestibule of inner ear

   - Oval window
   - Round window
   - Cochlea

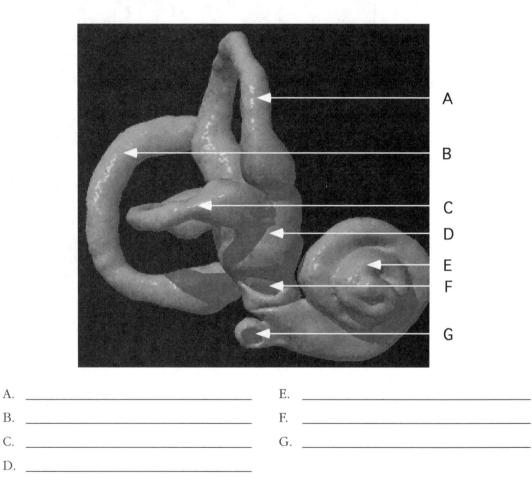

A. _____   E. _____

B. _____   F. _____

C. _____   G. _____

D. _____

Exercise 4.7 Structures of the Cochlea

1. Use the 3D Ear Structure List to label the structures in the following diagram:

   - Perilymph
   - Endolymph
   - Spiral organ

   - Tectorial membrane
   - Basilar membrane of cochlea
   - Spiral ganglion

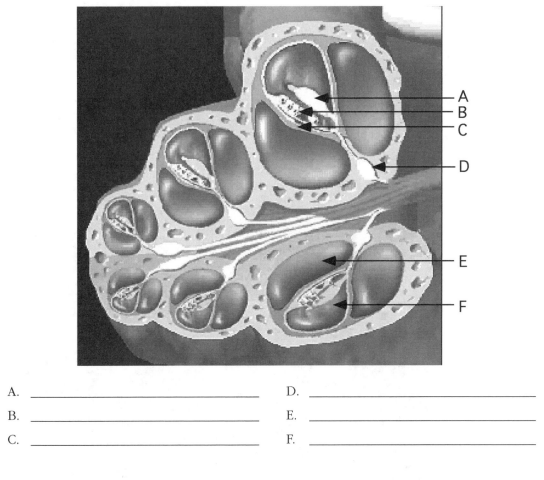

A. _____   D. _____

B. _____   E. _____

C. _____   F. _____

➤ *Close the 3D Ear window.*

2. Match the following structures with the appropriate ear region:

    I. Outer ear   II. Middle ear   III. Inner ear

    a. _____ Auricle (pinna)   f. _____ Incus

    b. _____ External auditory canal   g. _____ Stapes

    c. _____ Malleus   h. _____ Vestibule

    d. _____ Cochlea   i. _____ Auditory (eustachian) tube

    e. _____ Semicircular canals

Clinical Animation

To observe a clinical animation of hearing and the cochlea, do the following:

➤ *Open Content from the File Menu. Select Clinical Animations. Select Head and Neck from the Body Region drop-down menu. Select Hearing and the Cochlea. Click the Open button. Expand the window by clicking the Maximize button.*

➤ *A "Hearing and the Cochlea" window will open and begin to play. Below the animation window a text window can be scrolled through.*

➤ *After the animation finishes, close the "Hearing and the Cochlea" window.*

SMELL

Exercise 4.8 Nasal Cavity

➤ *Select Open Content from the File menu. Choose Atlas Anatomy.*

➤ *Select Head and Neck from the Body Region drop-down menu and Medial from the View Orientation drop-down menu. Select the "Olfactory Nerve in Nasal Cavity" thumbnail icon. Click Open.*

➤ *Expand the window.*

1. Label the lettered pins in the diagram that follows.

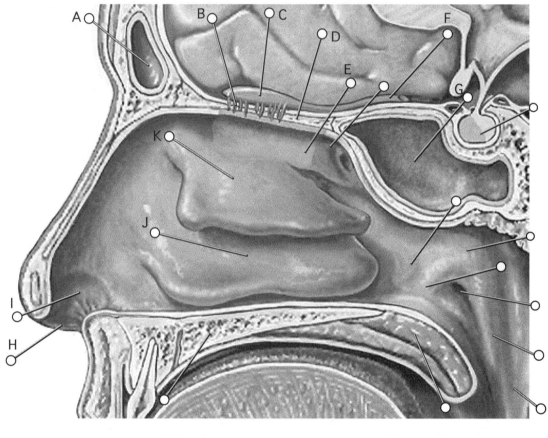

A. _____	G. _____
B. _____	H. _____
C. _____	I. _____
D. _____	J. _____
E. _____	K. _____
F. _____	

➤ *Close the Atlas Anatomy window.*

TASTE

Exercise 4.9 Tongue

➤ *Select Open Content from the File menu. Choose Atlas Anatomy.*

➤ *Select Head and Neck from the Body Region drop-down menu and Superior from the View Orientation drop-down menu. Select the "Surface of Tongue (Dorsal)" thumbnail icon. Click Open.*

➤ *Expand the window.*

1. Label the lettered pins on the diagram that follows. (You may need to adjust the diagram on the screen to find all of the pins.)

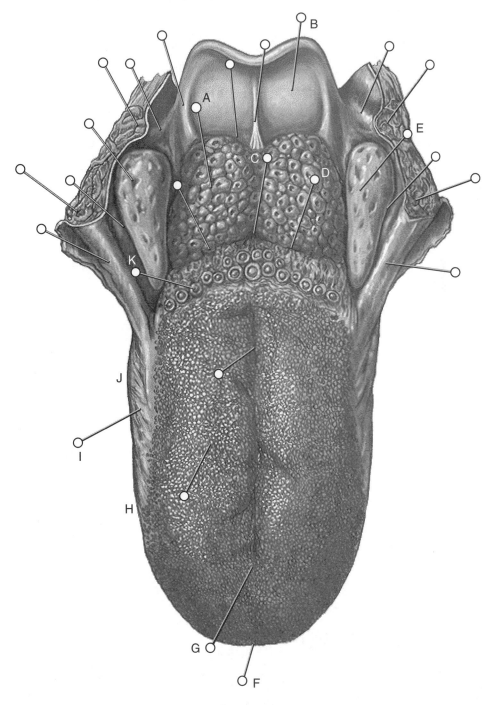

A. _____   G. _____

B. _____   H. _____

C. _____   I. _____

D. _____   J. _____

E. _____   K. _____

F. _____

**Beyond
A I A** 2. What is the function of the papillae?

_____

➤ *Close the Atlas Anatomy window.*

ENDOCRINE SYSTEM

5

STUDENT OBJECTIVES

OVERVIEW
- Review the anatomy of the endocrine system.

GLANDS OF THE HEAD AND NECK
- Describe the location and structure of the pituitary gland and the hypothalamus.
- List some important hormones produced by the hypothalamus and the pituitary gland.
- Describe the location and structure of the pineal gland.

- Describe the location and structure of the thymus, thyroid, and parathyroid glands.

GLANDS OF THE ABDOMEN
- Describe the structure and location of the adrenal glands.
- Describe the location and structure of the pancreas.
- Describe the location and structure of the testes in the male and the ovaries in the female.

GLANDS OF THE HEAD AND NECK
Exercise 5.1 Pituitary Gland and Hypothalamus

➤ *Open AIA by double-clicking Start Interactive Anatomy. Choose Dissectible Anatomy.*

➤ *Be sure that Male and Lateral are selected. Click Open.*

➤ *Adjust the image so that it is centered on the side of the head.*

➤ *Adjust the Layer Indicator to 288, and click the Highlight button in the toolbar.*

1. In this view, you are looking at a midsagittal section of the skull. Label the diagram that follows using the Identify tool.

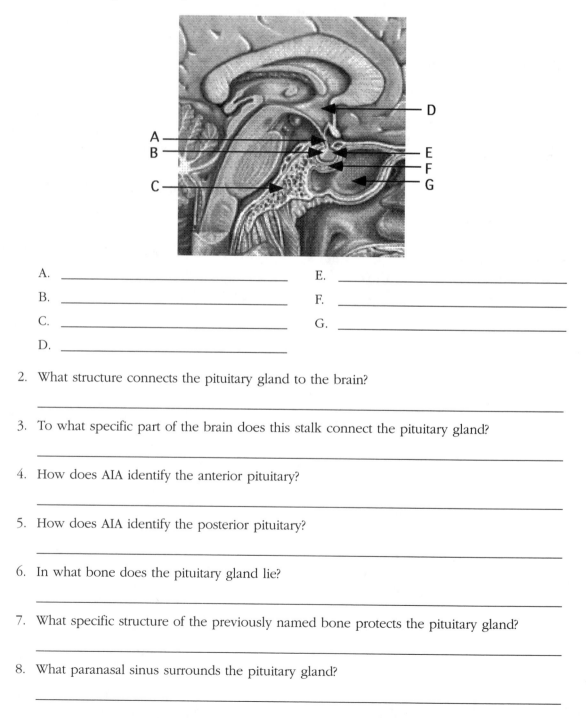

A. _____   E. _____

B. _____   F. _____

C. _____   G. _____

D. _____

2. What structure connects the pituitary gland to the brain?

_____

3. To what specific part of the brain does this stalk connect the pituitary gland?

_____

4. How does AIA identify the anterior pituitary?

_____

5. How does AIA identify the posterior pituitary?

_____

6. In what bone does the pituitary gland lie?

_____

7. What specific structure of the previously named bone protects the pituitary gland?

_____

8. What paranasal sinus surrounds the pituitary gland?

_____

**Beyond
A I A** 9. The anterior pituitary produces many tropic hormones. What does the term *tropic* mean?

_____

10. The pituitary stores two hormones made in the hypothalamus. What are they, and what are their functions?

    a. Hormone 1: _____

    b. Function: _____

    c. Hormone 2: _____

    d. Function: _____

    (Note: The hypothalamus is often called the master of the pituitary, indicating its direct control over the pituitary by various releasing or inhibiting hormones. It is the major link between the nervous and endocrine systems in the brain. It is not a part of the blood-brain barrier.)

Exercise 5.2 Pineal Gland

1. Label the diagram below using the Identify tool.

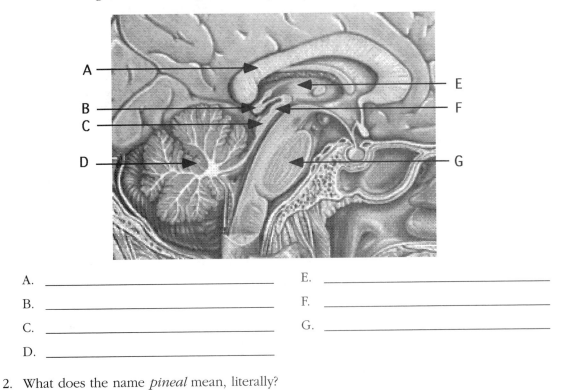

    A. _____   E. _____

    B. _____   F. _____

    C. _____   G. _____

    D. _____

**Beyond
A I A** 2. What does the name *pineal* mean, literally?

_____

3. The pineal gland produces melatonin. What is the function of melatonin?

_____

(Note: More recently, the pineal gland's production of serotonin [5-hydroxytryptamine] has been linked to an individual's behavior. Individuals who oversecrete serotonin become extremely passive. Conversely, individuals who have low levels of serotonin often exhibit aggressive and suicidal behavior. To combat this, medications that inhibit the reuptake of serotonin are often prescribed.)

Exercise 5.3 Thymus, Thyroid, and Parathyroid Glands

➤ *Choose Anterior from the View button drop-down menu.*

➤ *Adjust the image so that it is centered on the neck region, and set the Layer Indicator to 77.*

1. Label the diagram below using the Identify tool.

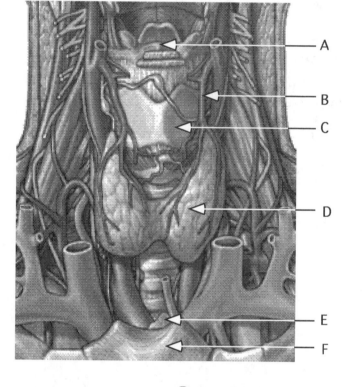

A. _____   D. _____

B. _____   E. _____

C. _____   F. _____

2. From what artery does the superior thyroid artery arise?

_____

Beyond A I A

3. What is the role of the thymus?

_____

(Note: The thymus is often identified as a remnant. This is because of its normal loss in size [atrophy] as one ages.)

4. The thyroid produces thyroxine. What is its function?

_____

(Note: Thyroxine is also known as T_4 because it contains four atoms of iodine. The thyroid also produces triiodothyronine, which contains three iodine atoms. This makes the thyroid the site in the body where iodine is concentrated. When working with radioactive iodine [such as I-125], researchers are checked for absorption of the radioactive iodine by placing a Geiger counter over the throat region.)

5. The thyroid produces calcitonin. What is its function?

_____

6. Although not visible in this view, the four parathyroid glands are located behind the thyroid (one superior and one inferior gland attached to each lateral thyroid lobe). What hormone do the parathyroids produce, and what is its function?

_____

Exercise 5.4 Glands of the Head and Neck

➤ *Close the Male Anterior window. Select Open Content from the File menu.*

➤ *Click Atlas Anatomy, and select Endocrine from the Body System drop-down menu and Lateral from the View Orientation drop-down menu. Select the "Glands of Head and the Neck (Lat)" thumbnail icon. Click Open.*

➤ *Expand the window. The image illustrates the major endocrine and exocrine glands of the head and neck.*

1. Label the lettered pins in the following diagram.

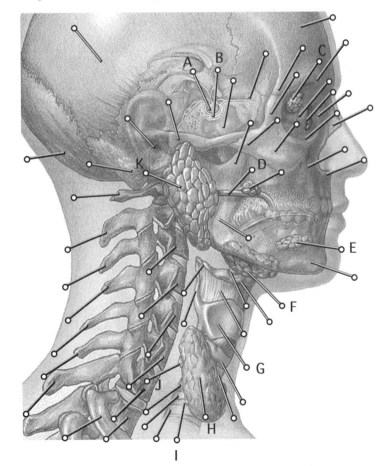

A. _____   G. _____

B. _____   H. _____

C. _____   I. _____

D. _____   J. _____

E. _____   K. _____

F. _____

Beyond
A I A

2. By what other name is the hypophysial fossa of the sphenoid bone known?

   _____

3. What is the literal meaning of *sella turcica*?

   _____

4. What three glands constitute the salivary glands?

   _____

5. By what other name is the parotid duct known?

   _____

6. What brain region controls the release of hormones by the pituitary gland?

   _____

7. In terms of size, where does the thyroid cartilage rank in the architecture of the larynx?

   _____

8. What is the literal meaning of *hyoid*?

   _____

   (Note: The hyoid bone is characteristically fractured in homicides involving strangulation.)

9. By what name are secretions of the lacrimal gland known, and what is the function of these secretions?

   _____

GLANDS OF THE ABDOMEN

Exercise 5.5 Adrenal (Suprarenal) Glands

➤ *Close the "Glands of Head and Neck (Lat)" window, and reopen the Dissectible Anatomy window. Be sure that Male Anterior is selected, and click Open. Expand the window.*

➤ *Enlarge and adjust the image so that it is centered over the abdominal region, and adjust the Layer Indicator to 232.*

1. Label the diagram below using the Identify tool.

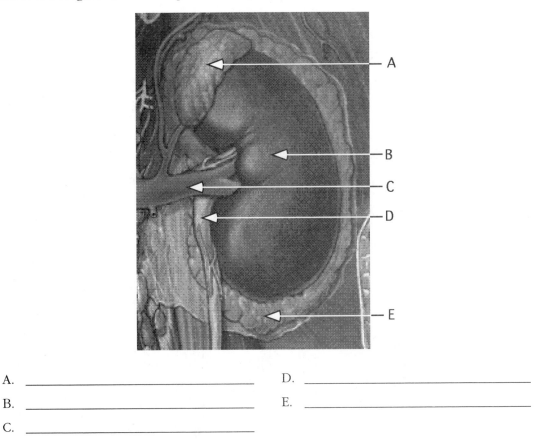

A. _____   D. _____

B. _____   E. _____

C. _____

2. What is the name of the outer region of the adrenal glands that makes up the bulk of the adrenal gland and surrounds the inner region?

_____

3. What is the name of the inner region of the adrenal glands?

_____

4. The adrenal cortex produces mineralocorticoids. About 95% of the mineralocorticoids produced is in the form of aldosterone. How does aldosterone affect blood volume?

_____

_____

5. The adrenal medulla produces epinephrine (adrenaline) and norepinephrine (noradrenaline). Epinephrine accounts for 80% of the secretions of the adrenal medulla. Why is epinephrine considered a sympathomimetic?

_____

_____

Exercise 5.6 Pancreas

➤ *Set the Layer Indicator to 214.*

1. Label the diagram below using the Identify tool.

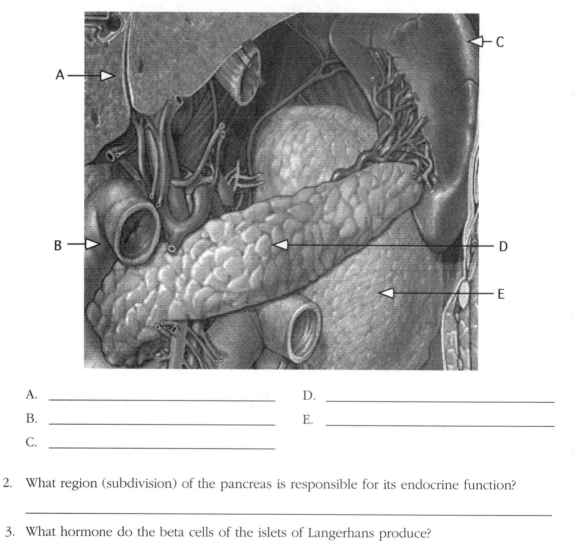

A. _____   D. _____

B. _____   E. _____

C. _____

**Beyond
A I A**
2. What region (subdivision) of the pancreas is responsible for its endocrine function?

_____

3. What hormone do the beta cells of the islets of Langerhans produce?

_____

4. What hormone do the alpha cells of the islets of Langerhans produce?

_____

5. What is the role of insulin?

_____

6. What is the role of glucagon?

_____

7. What disease is associated with a lack of insulin?

_____

(Note: The pancreas is considered to be a mixed gland, in that it has both endocrine
[production of insulin and glucagon] and exocrine [production of digestive enzymes]
functions.)

Exercise 5.7 Testes

➢ *Adjust the image so that the pubic region is centered, and set the Layer Indicator to 180.*

1. Use the Identify tool to label the diagram below.

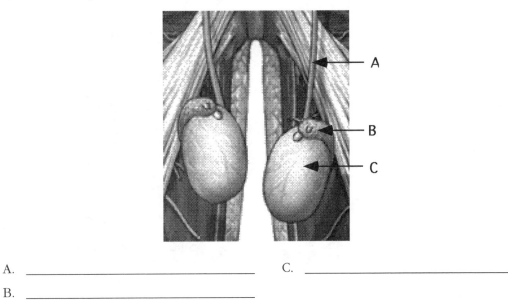

A. _____   C. _____

B. _____

**Beyond
A I A**

2. What hormone is produced by the testes?

_____

3. What is the function of the ductus deferens?

_____

4. What is the function of the epididymis?

_____

5. What is the function of testosterone?

_____

Exercise 5.8 Ovaries

➤ *Choose Female from the Gender button drop-down menu. Set the Layer Indicator to 226.*

1. Label the diagram below.

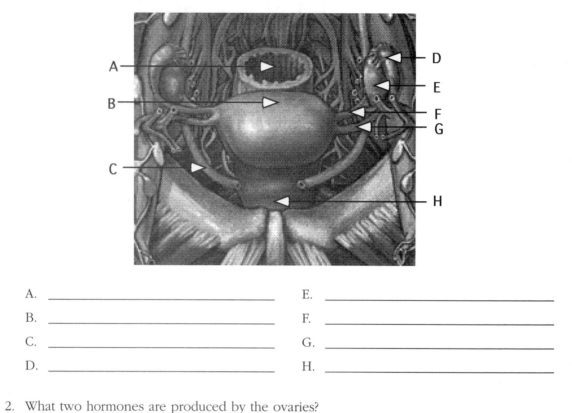

A. _____   E. _____

B. _____   F. _____

C. _____   G. _____

D. _____   H. _____

Beyond A I A

2. What two hormones are produced by the ovaries?

   _____

3. What is the function of estrogen?

   _____

4. What is the function of progesterone?

   _____

Clinical Animation

To observe a clinical animation of endocrine glands, do the following:

➤ *Select Open Content from the File Menu. Select Clinical Animations. Select Head and Neck from the Body Region drop-down menu. Select "Endocrine glands—general overview." Click the Open button. Expand the window by clicking the Maximize button.*

➤ *An "Endocrine glands—general overview" window will open and begin to play. Below the animation window a text window can be scrolled through.*

➤ *After the animation finishes, close the "Endocrine glands—general overview" window.*

CARDIOVASCULAR SYSTEM

6

STUDENT OBJECTIVES

OVERVIEW

- Review the anatomy of the heart and the arterial and venous systems.

HEART

- Describe the components of the pericardium, the visceral and parietal portions.

- Describe the external anatomy of the heart and its great vessels.

- Describe the location of the heart's coronary blood vessels.

- Describe the heart's internal anatomy, including its chambers, atrioventricular valves, semilunar valves, papillary muscles, and chordae tendineae.

ARTERIES

- Identify the major arteries that supply blood to the head and neck, upper and lower extremities, thorax, abdomen, and pelvis.

- Describe the formation of a special arterial supply to the brain, the circle of Willis.

- Identify the major arterial branches of the ascending aorta, aortic arch, thoracic aorta, and abdominal aorta.

- Describe the anatomical differences between an artery and a vein.

VEINS

- Identify the major veins responsible for returning venous blood to the heart from the upper and lower extremities, thorax, abdomen, hips, scalp, head, neck, face, and shoulder.

HEART

Exercise 6.1 Superficial View of the Heart

➤ *Open AIA by double-clicking Start Interactive Anatomy. Select Dissectible Anatomy.*

➤ *Be sure that Male (or Female) and Anterior are selected. Click Open. Expand the window.*

➤ *Magnify the image.*

➤ *Adjust the primary window so that it is centered on the chest.*

➤ *Adjust the Layer Indicator to 173, and click the Highlight button in the toolbar.*

➤ *In this view, the anterior chest and abdominal wall have been removed, as well as the anterior fibrous (parietal) pericardium, so that the heart and its great vessels can be seen.*

Beyond AIA 1. Describe the mediastinum and its boundaries.

_____

2. Use the Identify tool to label the structures in the diagram below.

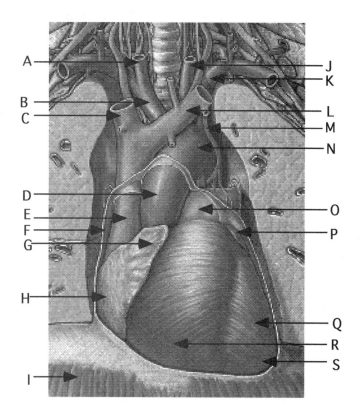

A. _____   K. _____

B. _____   L. _____

C. _____   M. _____

D. _____   N. _____

E. _____   O. _____

F. _____   P. _____

G. _____   Q. _____

H. _____   R. _____

I. _____   S. _____

J. _____

➢ *Select the Normal button to restore full color.*

3. What is the name of the small conical muscular pouch that projects to the left from the right atrium and overlaps the ascending aorta?

   _____

4. What is the name of the major blood vessel that delivers deoxygenated blood from the superior half of the body into the right atrium of the heart?

   _____

5. Describe the aortic arch.

   _____

   _____

6. What is the name of the large muscular partition (in the preceding figure) between the thoracic and abdominal cavities?

   _____

7. What structure (in the preceding figure) represents the parasympathetic innervation of the heart?

   _____

Beyond A I A

8. What is the role of the right atrium?

   _____

9. What is the function of the aorta?

   _____

Exercise 6.2 Coronary Blood Vessels

➤ *Move the Layer Indicator to 171.*

1. What is the name of the artery passing inferiorly from the base of the ascending aorta and following the course of the coronary sulcus between the right auricle and the right ventricle?

_____

(Note: Typically, the artery that you just identified supplies oxygenated blood to the right atrium, the right ventricle, and the interatrial septum, including the sinoatrial [SA] and atrioventricular [AV] nodes. It also supplies a variable part of the left atrium and the left ventricle.)

2. What is the name of the artery running down the course of the anterior interventricular sulcus?

_____

(Note: The anterior interventricular artery represents one of the two terminal branches of the left coronary artery, the other being the circumflex artery. The anterior interventricular artery supplies most of the left ventricle, some of the right ventricle, and the interventricular septum, including the AV bundle. The anterior interventricular artery is also known as the left anterior descending [LAD] artery. Atherosclerosis of the LAD artery can lead to ischemia [lack of blood flow to a muscle], angina [cramping of the muscle], and possible infarction [cell death caused by a lack of blood nutrients]. This can have drastic effects on the functioning of the heart muscle's ventricles, particularly the left ventricle, and often causes fatal heart attacks, leading cardiologists to refer to the LAD as the "widow maker.")

3. What are the two veins called that cross over the superior part of the right coronary artery onto the anterior surface of the right ventricle?

_____

4. What is the name of the vein passing along the inferior portion of the right ventricle?

_____

5. What is the name of the vein beginning at the apex of the heart and ascending in the anterior interventricular groove with the anterior interventricular artery?

_____

(Note: The vein you just identified is a main tributary flowing into the largest vein of the heart, the coronary sinus. The coronary sinus drains all the venous blood from the heart except that carried by the anterior cardiac veins.)

Clinical Animation

Three procedures used by cardiologists to treat coronary arteries that become occluded (clogged) due to atherosclerosis are percutaneous transluminal coronary angioplasty (PTCA), directional coronary atherectomy (DCA), and coronary artery bypass grafting (CABG). To observe a clinical animation of each of these techniques, do the following:

➤ *Select Open Content from the File Menu. Select Clinical Animations. Select All from the Body Region drop-down menu and Cardiovascular from the Body System drop-down menu.*

➤ Select the animation desired: Percutaneous transluminal coronary angioplasty (PTCA), Directional coronary atherectomy (DCA), or Coronary artery bypass grafting (CABG). Click the Open button. Expand the window by clicking the Maximize button.

➤ A window will open and begin to play. After the animation finishes, close the window. All three techniques can be viewed by repeating the above directions.

Exercise 6.3 Anterior View of the Heart

➤ Adjust the Depth bar to 174. In this view, parts of the anterior walls of the right atrium, right ventricle, and left atrium have been removed.

1. Click the Highlight button in the Tool palette. Label the following diagram:

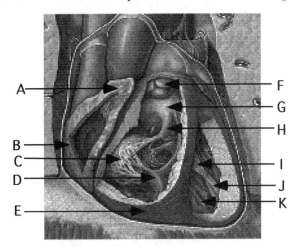

A. _____	G. _____
B. _____	H. _____
C. _____	I. _____
D. _____	J. _____
E. _____	K. _____
F. _____	

Beyond A I A

2. What are the names of the four chambers of the heart?

_____

3. As you view the heart-cut section image, which ventricle, right or left, has the thicker walls?

_____

4. What is the physiological significance of the difference in thickness?

_____

5. Name the two AV valves.

_____

6. Describe the location of the two AV valves.

_____

7. Name the two semilunar valves.

   _____

8. Describe the location of the two semilunar valves.

   _____

Beyond
A I A

9. What is the function of the papillary muscles of the heart?

   _____

10. What are *trabeculae carneae*?

    _____

11. What is pulmonary valve stenosis?

    _____

Exercise 6.4 Lateral View of the Heart

➤ *Select Lateral from the View button drop-down menu. Center the image over the chest and adjust the Layer Indicator to 219.*

1. Use the Identify tool to label the following diagram.

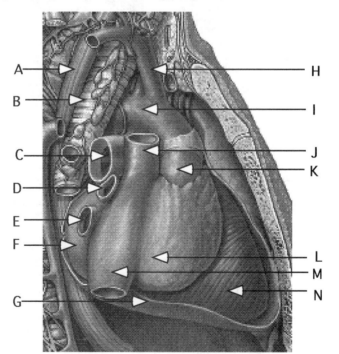

A. _____	H. _____
B. _____	I. _____
C. _____	J. _____
D. _____	K. _____
E. _____	L. _____
F. _____	M. _____
G. _____	N. _____

Exercise 6.5 Sagittal Section of the Heart

➤ *Adjust the Layer Indicator to 220.*

1. Use the Identify tool to label the following diagram.

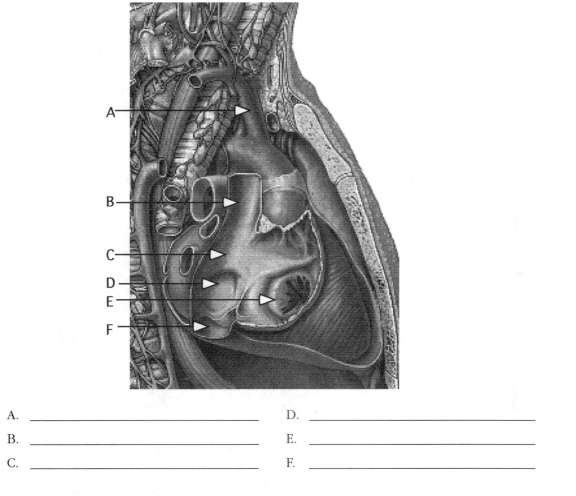

A. _____   D. _____

B. _____   E. _____

C. _____   F. _____

2. What is the significance of the red color of the right superior and inferior pulmonary veins in this view?

_____

3. What is the significance of the blue color of the right pulmonary artery in this view?

_____

Beyond A I A

4. What fetal structure gives rise to the fossa ovalis?

_____

5. What does *brachiocephalic* mean, literally?

_____

➤ *Close the Lateral window.*

Exercise 6.6 Sagittal Section of the Left Ventricle

➤ *Select Open from the File menu and choose 3D Anatomy. Select 3D Heart. Expand the window.*

➤ *Choose the Anterior cusp of mitral (Left AV) valve from the Structure List. An image of the heart appears showing a sagittal section through the left ventricle.*

1. Label the following structures in the figure below. (Note: The first three structures can be identified by clicking and holding the Structure List, located just above the heart image, and selecting the appropriate structure name from the Structure List. However, "Chordae tendineae" and "Myocardium of the left ventricle" are not identified by AIA. It may be necessary to use reference materials to label these structures.)

- Anterior cusp of mitral (left AV) valve
- Trabeculae carneae
- Anterior papillary muscle—left ventricle
- Posterior papillary muscle
- Chordae tendineae
- Myocardium of left ventricle

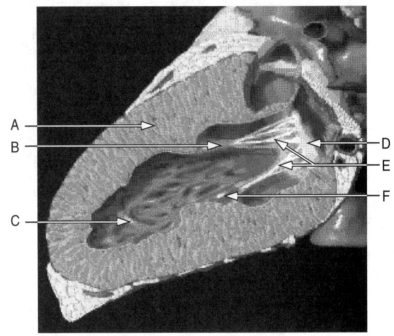

A. _____   D. _____

B. _____   E. _____

C. _____   F. _____

Beyond A I A 2. By what other name is the AV mitral valve known?

_____

3. What is the significance of the other name of the AV mitral valve?

_____

Exercise 6.7 Sagittal Section of the Right Ventricle

➤ *Select Anterior cusp of the tricuspid (right) valve from the Structure List.*

1. Label the following structures on the diagram below. (Note: "Myocardium of right ventricle" is not identified by AIA in this image. It may be necessary to use reference materials to label this structure.)

   - Anterior papillary muscle—right ventricle
   - Anterior cusp of tricuspid (right AV) valve
   - Septomarginal trabecula
   - Posterior cusp of tricuspid (right AV) valve
   - Myocardium of right ventricle

   - Right cusp of pulmonary valve
   - Conus arteriosus
   - Auricle of right atrium
   - Septal cusp of the tricuspid (right AV) valve—anterior

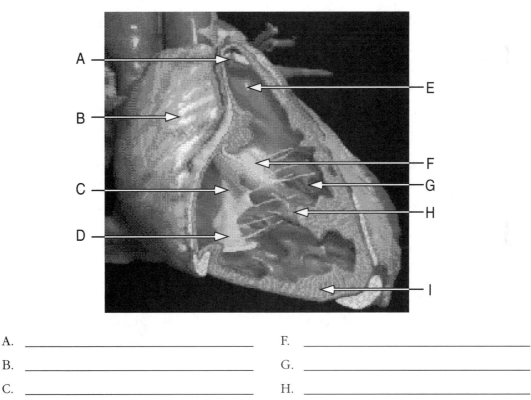

A. _____   F. _____

B. _____   G. _____

C. _____   H. _____

D. _____   I. _____

E. _____

Beyond A I A

2. What do the letters AV represent when describing the mitral and/or tricuspid valves, and how do the letters AV indicate heart valve location?

   _____

3. What specific role does the tricuspid valve serve?

   _____

4. Septal defects are the most common congenital heart disease, representing 30% to 40% of all clinically recognized cases. Interventricular septal defects are more common than interatrial septal defects.

   a. What is an interatrial septal defect?

   _____

   b. What is a ventricular septal defect (VSD)?

   _____

   c. What are the consequences of a VSD?

   _____

   _____

Exercise 6.8 Electrical Conduction System of the Heart

Clinical Animation

To observe a clinical animation of the cardiac conduction system, do the following:

➤ *Select Open Content from the File Menu. Select Clinical Animations. Select All from the Body Region drop-down menu and Cardiovascular from the Body System drop-down menu. Select "Cardiac conduction system" and click open.*

➤ *A clinical animation of the cardiac conduction system will begin to play. Once completed, close the clinical animation window.*

1. The following diagram is not presented in AIA. Use reference materials to label the diagram that follows.

   - Superior vena cava
   - Left atrium
   - Right ventricle
   - Sinoatrial (SA) node
   - AV bundle
   - Aortic arch
   - Right and left bundle branches

   - Left ventricle
   - Inferior vena cava
   - Ascending aorta
   - Right atrium
   - Atrioventricular (AV) node
   - Purkinje fiber

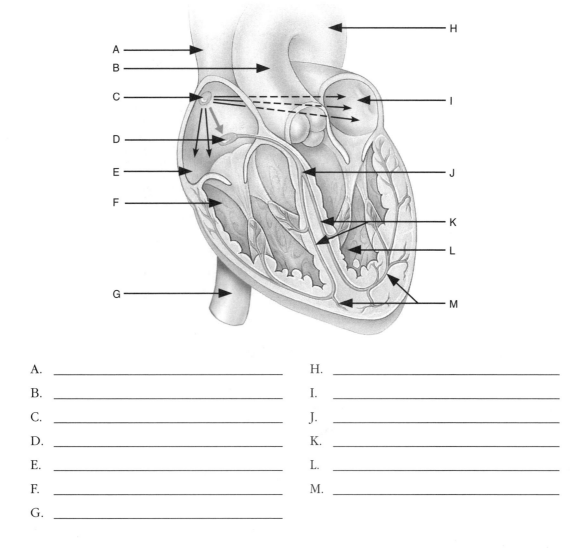

A. _____   H. _____

B. _____   I. _____

C. _____   J. _____

D. _____   K. _____

E. _____   L. _____

F. _____   M. _____

G. _____

ARTERIES

Exercise 6.9   Arteries of the Head and Neck

➤ *Close 3D Anatomy and open Atlas Anatomy.*

➤ *Select Head and Neck from the Body Region drop-down menu and Cardiovascular from the Body System drop-down menu. Select the "Arteries of Head & Neck" thumbnail icon. Click Open.*

➤ *Expand the window.*

1. Label the following diagram.

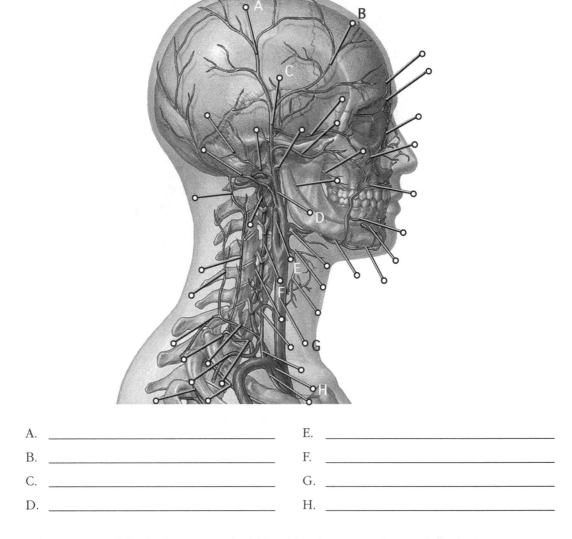

A. _____   E. _____

B. _____   F. _____

C. _____   G. _____

D. _____   H. _____

Beyond
A I A

2. What regions of the body are supplied blood by the external carotid artery?

_____

3. What is the significance of transverse foramina in cervical vertebrae?

_____

4. Through what cranial structure does the carotid artery pass as it enters the brain?

_____

Exercise 6.10 The Cerebral Arterial Circle (Circle of Willis)

➤ *Close the window labeled Arteries of Head & Neck.*

➤ *Select Open Content from the File menu and choose Atlas Anatomy.*

➤ *Select Head and Neck from the Body Region drop-down menu, Cardiovascular from the Body System drop-down menu, and Inferior from the View menu.*

➤ *Select the "Cerebral Arterial Circle (Inf)" thumbnail icon. Click Open.*

➤ *Expand the window.*

➤ *The image shows the base of the brain with the cranial nerves and blood vessel supply. Also visible is the special circular anastomosis of the four arteries that supply blood to the brain, known as the circle of Willis (cerebral arterial circle).*

1. Label the following diagram.

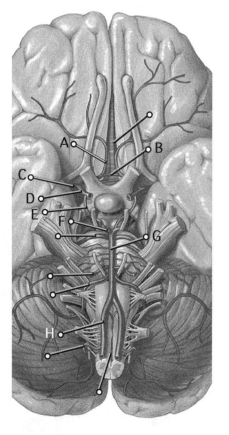

A. _____	E. _____
B. _____	F. _____
C. _____	G. _____
D. _____	H. _____

2. The two vertebral arteries merge to form what artery?

    _____

3. The circle of Willis encircles cranial nerve II (optic nerve) and what brain structure?

    _____

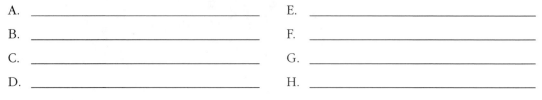

Beyond
A I A

4. What is an anastomosis?

    _____

5. What is the physiological importance of an anastomosis?

    _____

Exercise 6.11 The Cerebral Arterial Circle (Circle of Willis)

➤ *Close the Atlas Anatomy window.*

➤ *Select Open Content from the File menu. Click 3D Anatomy and select Brain. Click Open.*

➤ *Maximize the view on your screen.*

1. Using the Structure List for Brain at the top of your screen, scroll down the list and identify the following structures on the diagram below. (Note: The structures are listed in alphabetical order.)

- Anterior cerebral artery
- Anterior communicating artery
- Basilar artery
- Internal carotid artery (left)

- Internal carotid artery (right)
- Posterior cerebral artery
- Posterior communicating artery
- Vertebral artery (right)

Rostral / Anterior end

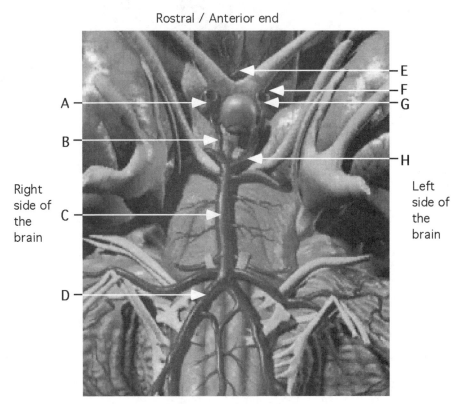

Caudal / Posterior end

Ventral / Inferior View

A. _____   E. _____

B. _____   F. _____

C. _____   G. _____

D. _____   H. _____

Exercise 6.12 Arterial Branches of the Arch of the Aorta

➤ *Close the 3D Brain window.*

➤ *Select Open Content from the File menu. Select Dissectible Anatomy, Male, and Anterior, and click Open.*

➤ *Enlarge the window.*

➤ *Adjust the Layer Indicator to 246.*

1. Magnify and adjust the image on your screen to match the following figure. Label the following figure.

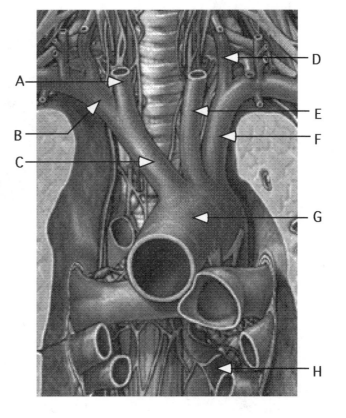

A. _____	E. _____
B. _____	F. _____
C. _____	G. _____
D. _____	H. _____

2. The right common carotid artery is formed at the bifurcation (splitting or fork) of the first branch of the aortic arch. What is the first branch?

_____

3. What is the second branch of the aortic arch?

_____

Exercise 6.13 Arteries of the Arm

➤ *Close the Dissectible Anatomy window.*

➤ *Select Open Content from the File menu.*

➤ *Select Atlas Anatomy.*

➤ *Select Upper Limb from the Body Region menu, Cardiovascular from the Body System menu, and Anterior from the View Orientation menu.*

➤ *Select the "Arteries of the Upper Limb (Ant)" thumbnail icon. Click Open.*

➤ *Expand the window.*

1. Label the following diagram.

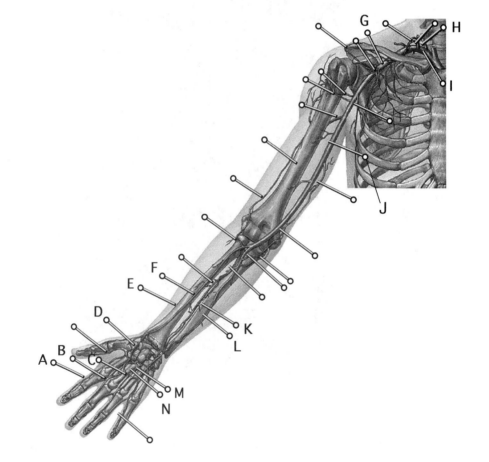

A. _____	H. _____
B. _____	I. _____
C. _____	J. _____
D. _____	K. _____
E. _____	L. _____
F. _____	M. _____
G. _____	N. _____

2. From what blood vessel does the axillary artery receive its blood?

   _____

3. What is the name of the large artery (a continuation of the axillary artery) running on the medial side of the arm, which represents the principal arterial supply to the arm?

   _____

   (Note: Arterial blood pressure levels are routinely measured using a sphygmomanometer, a device consisting of an inflatable cuff and a mercury manometer. The cuff is placed around the arm and inflated to compress the brachial artery against the humerus.)

   (Note: Pulsations of the brachial artery, in addition to skeletal muscle contractions, assist blood movement through the venous network.)

4. What is the name of the large artery, representing one of the two terminal branches of the brachial artery, running along the lateral portion of the forearm?

   _____

5. What is the name of the larger of the two terminal branches of the brachial artery, which runs along the medial portion of the forearm?

   _____

**Beyond
A I A**

6. What is the derivation of *radial artery?*

   _____

   (Note: The radial artery is commonly used to take blood pulse measurements near the wrist.)

Exercise 6.14 Abdominal Aorta and Its Branches

➤ *Close the Atlas Anatomy window entitled "Arteries of Upper Limb (Ant)."*
➤ *Open Dissectible Anatomy and click Male and Anterior. Click Open.*
➤ *Expand the window.*
➤ *Adjust the Layer Indicator to 240.*

1. Adjust the image to match the diagram that follows, and label the diagram.

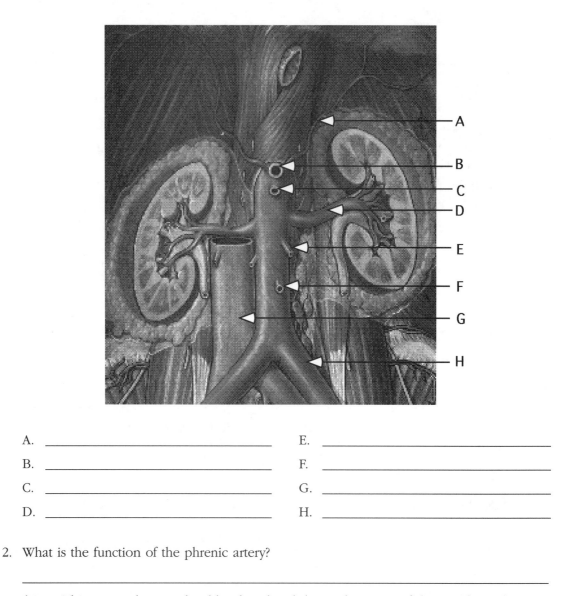

A. _____   E. _____

B. _____   F. _____

C. _____   G. _____

D. _____   H. _____

Beyond
A I A
2. What is the function of the phrenic artery?

_____

(Note: This artery also supplies blood to the abdominal portion of the esophagus.)

3. What artery in a female would correspond to the testicular artery?

_____

4. The celiac artery is a short, wide vessel, extending about 1 cm anteriorly and immediately dividing into three branches.

a. What branch supplies blood to the spleen?

_____

(Note: This is the largest branch of the celiac artery.)

b. What branch supplies blood to the stomach?

_____

(Note: This is the smallest branch of the celiac artery. The stomach actually receives blood from all three branches of the celiac artery.)

c. What branch supplies blood to the liver?

_____

(Note: The liver has a double blood supply, receiving 30% oxygenated blood from the branch of the celiac artery that you just identified and 70% venous blood from the portal vein, containing the products of digestion absorbed from the gastrointestinal tract.)

5. What organs does the superior mesenteric artery supply?

_____

(Note: The inferior mesenteric artery supplies the remaining portion of the gastrointestinal tract distal to the left colic flexure.)

Exercise 6.15 Blood Vessels of the Upper Leg and Hips

1. Adjust the image to match the diagram below, and then label the diagram.

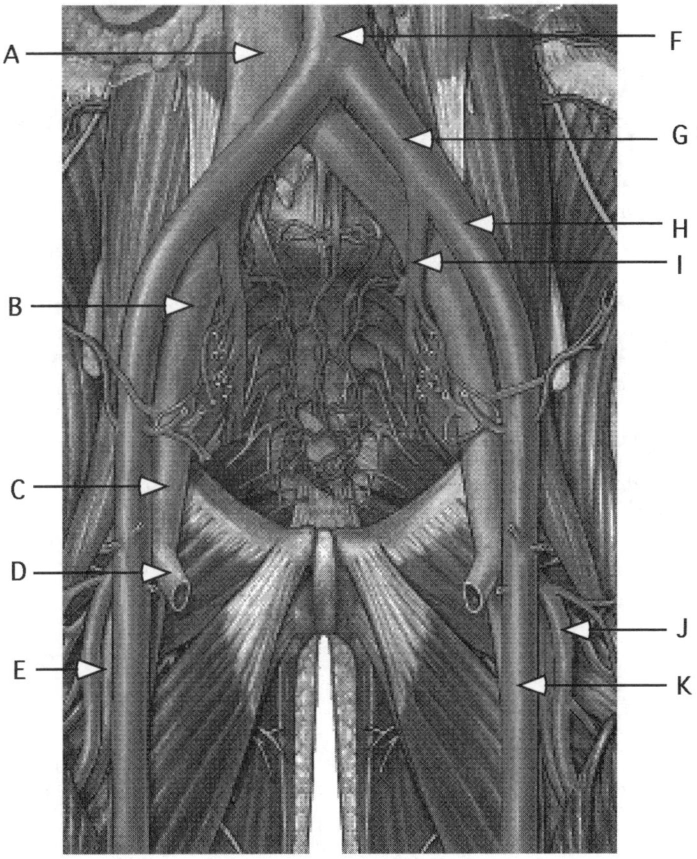

A. _____   G. _____

B. _____   H. _____

C. _____   I. _____

D. _____   J. _____

E. _____   K. _____

F. _____

2. What two arteries represent the terminal branches of the abdominal aorta?

_____

3. What are the two terminal branches of the common iliac artery?

_____

(Note: The internal iliac artery supplies most of the blood to the pelvic viscera.)

4. Into what vein does the great saphenous vein drain?

_____

5. What large artery, representing the continuation of the external iliac artery, provides the chief arterial supply to the lower limb?

_____

(Note: The femoral artery can be easily exposed in the femoral triangle and is commonly the site of insertion of catheters when cardiac angiography is performed. In addition, the femoral artery lies superficially within the femoral triangle, making it particularly vulnerable to deep cuts and puncture by gunshot wounds.)

Exercise 6.16 Arteries of the Lower Appendages

➤ _Close the Male Anterior window._

➤ _Select Open Content from the File Menu. Choose Atlas Anatomy._

➤ _Select Lower Limb from the Region menu, Cardiovascular from the System menu, and Posterior from the View menu._

➤ _Select the "Arteries of the Lower Limb (Post)" thumbnail icon. Click Open._

➤ _Select Open Content from the File menu again._

➤ _Select Lower Limb from the Region menu, Cardiovascular from the System menu, and Anterior from the View menu._

➤ _Select the "Arteries of the Lower Limb (Ant)" thumbnail icon. Click Open._

➤ _Select Tile Vertically from the Window menu._

1. Adjust the images in both windows so that they match the diagram below, and then place the correct names of the labeled letters in the appropriate spaces below the diagrams. (Note: Corresponding letters appearing on both the anterior and posterior diagrams are pointing to the same structures.)

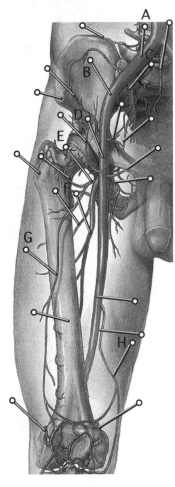

Anterior

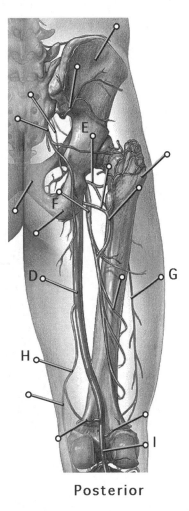

Posterior

A. _____   F. _____

B. _____   G. _____

C. _____   H. _____

D. _____   I. _____

E. _____

2. Adjust the images in both windows to match the following diagrams and then label the diagrams. (Note: Corresponding letters appearing on both the anterior and posterior diagrams are pointing to the same structures.)

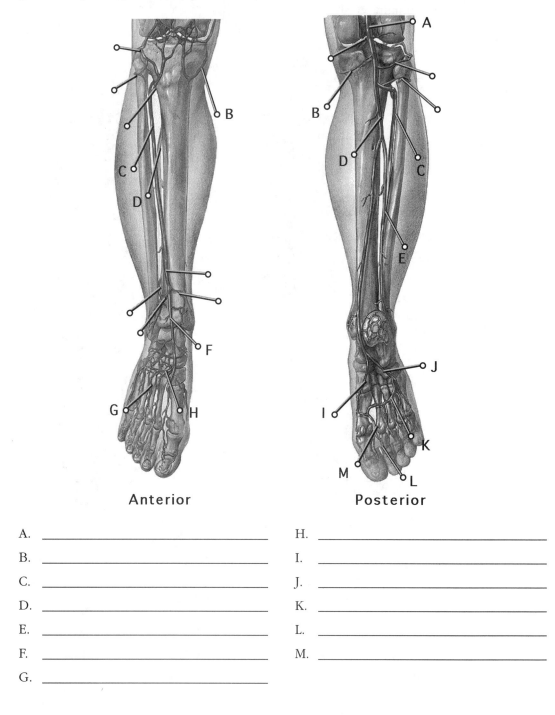

Anterior   Posterior

A. _____   H. _____

B. _____   I. _____

C. _____   J. _____

D. _____   K. _____

E. _____   L. _____

F. _____   M. _____

G. _____

3. What is the largest artery of the posterior knee region?

_____

4. Into what two arteries does the popliteal artery divide?

_____

5. Which terminal branch of the popliteal artery is the larger?

_____

(Note: You can palpate [feel] your pulse in this artery by palpating about halfway between the posterior surface of your medial malleolus and the medial border of your Achilles tendon.)

Beyond
A I A

6. Some of the arteries identified in the previous diagrams are genicular arteries. What does the prefix *genu* mean, and how is this term related to its arterial location?

_____

VEINS

Exercise 6.17 Veins of the Upper Extremity

➤ *Close the Atlas Anatomy windows. Select Open Content from the File menu and choose Dissectible Anatomy.*

➤ *Be sure that Male and Anterior are selected. Click Open. Expand the window.*

➤ *Adjust so that the right arm and elbow region are visible.*

➤ *Adjust the Layer Indicator to 5.*

1. Use the Identify tool to label the following diagram.

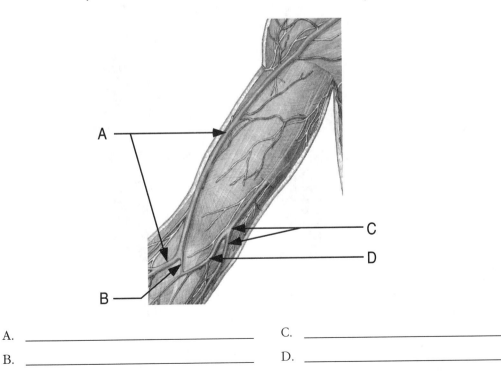

A. _____   C. _____

B. _____   D. _____

2. What is the large, medial, superficial vein passing up the arm?

   _____

   (Note: This vein is often visible through the skin.)

3. What is the large, lateral, superficial vein running along the anterolateral surface of the biceps brachii muscle up on the chest region?

   _____

   (Note: This vein becomes prominent on the biceps when one is lifting weights.)

4. Name the medial extension of the cephalic vein that joins with the median vein in front of the elbow (cubital fossa).

   _____

   (Note: The vein just identified forms the communication of the basilic and cephalic veins in the cubital fossa and, because of its prominence in the cubital fossa, is often used for taking blood [venipuncture].)

Exercise 6.18 Veins of the Lower Extremity

➤ _Select Posterior from the View button drop-down menu. Adjust the Layer Indicator to 3._

1. Adjust the image on your screen to match the diagram below, and label the diagram.

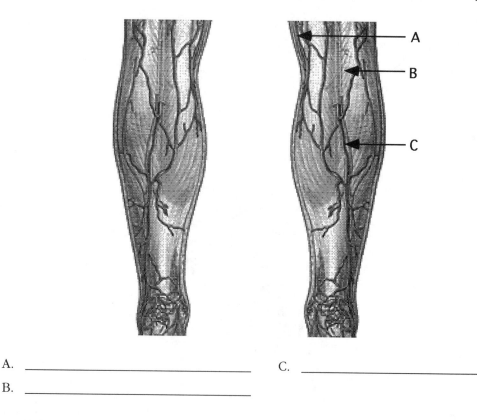

A. _____   C. _____

B. _____

2. What is the name of the long vein located in the groove between the two bellies of the gastrocnemius muscle?

   _____

3. Does the small saphenous vein begin on the posterior of the lateral malleolus of the fibula or the medial malleolus of the tibia? (Hint: Click the Highlight Mode icon of the toolbar to follow the course of this vein.)

_____

(Note: Valves of the lower limb veins can become dilated so that their cusps do not close. When this happens, contractions of calf muscles, which normally propel blood upward through the leg and thigh, cause a reverse flow. As a result, veins become twisted and dilated, resulting in varicose veins.)

Beyond A I A 4. One of the answers on the previous diagram above was *crural fascia*. What is meant by this term?

_____

➢ *Set the Layer Indicator to 129.*

5. Adjust the image on your screen to match the diagram below, and label the diagram.

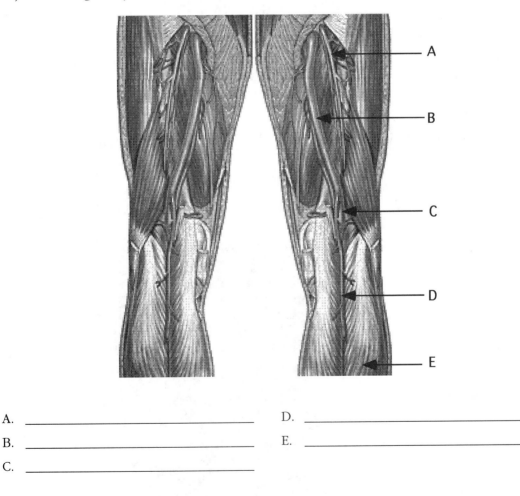

A. _____   D. _____

B. _____   E. _____

C. _____

Beyond A I A 6. Blood moves from the small saphenous vein into what vein?

_____

7. What does the name *popliteal* mean, literally?

_____

➤ *Select Anterior from the View button drop-down menu. Adjust the image so that it is centered on the knees.*

➤ *Adjust the Layer Indicator to 5. Select the Highlight button and locate the great saphenous vein (the large, long, blue vein ascending from the foot on the medial side of the leg, knee, and thigh up to the groin region). (Note: The great saphenous vein is the longest vein in the body, draining venous blood from the dorsum and sole of the foot.)*

➤ *To see the great saphenous vein in a medial view, select Medial from the View button drop-down menu. Set the Layer Indicator of the Male Medial window to 4.*

8. Does the great saphenous vein pass anteriorly or posteriorly to the majority of the tibia?

_____

9. Does the great saphenous vein pass anteriorly or posteriorly to the medial malleolus of the tibia?

_____

10. What is the name of the muscle that is located posteriorly to the great saphenous vein in the upper portion of the lower leg?

_____

➤ *Select Anterior from the View button menu.*

Exercise 6.19 Veins of the Hips and Abdomen

➤ *Adjust the image to the abdominal area.*

1. Adjust the Layer Indicator to 238. Label the following diagram. (Image adjustments may be necessary.)

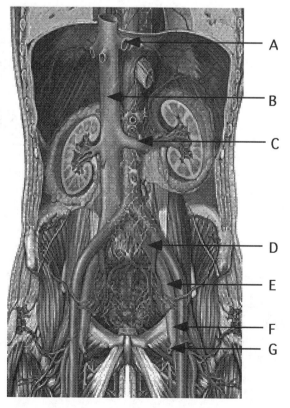

A

B

C

D

E

F

G

A. _____   E. _____

B. _____   F. _____

C. _____   G. _____

D. _____

2. Into what larger vein does the great saphenous vein drain?

   _____

Beyond A I A 3. What organ is drained by the hepatic veins?

   _____

Exercise 6.20 Veins of the Scalp, Face, and Neck

➤ *Adjust the image so that it is centered on the throat. Adjust the Layer Indicator to 30.*

1. Identify the following structures on the diagram below:

   - External jugular vein
   - Sternal head of sternocleidomastoid muscle
   - Clavicular head of sternocleidomastoid muscle
   - Anterior jugular vein
   - Manubrium
   - Clavicle

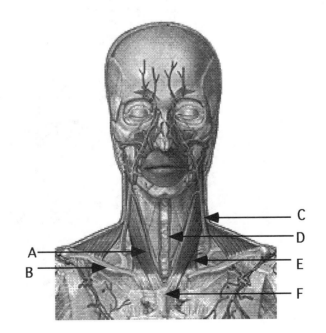

A. _____   D. _____

B. _____   E. _____

C. _____   F. _____

2. Name the two large veins running inferolaterally from the angles of the mandible down onto the shoulder region.

_____

3. What strap muscle of the neck does the external jugular vein cross as it descends to the shoulder?

_____

4. What is the name of the two parallel veins running inferiorly from the base of the chin down to the superior sternal region?

_____

Beyond AIA 5. From what areas of the body does the external jugular vein drain blood?

_____

Exercise 6.21 Major Veins of the Head, Neck, and Shoulder

➤ _Set the Layer Indicator to 76._

1. Identify the following structures on the diagram below:

- External jugular vein
- Internal jugular vein
- Axillary vein

- Basilic vein
- Right brachiocephalic vein
- Subclavian vein

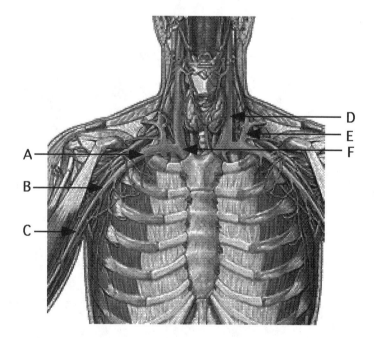

A. _____   D. _____

B. _____   E. _____

C. _____   F. _____

2. What is the name of the large vein that represents a continuation of the basilic vein and that ends between the first and second ribs?

_____

(Note: Wounds in the axilla [armpit] often involve the vein that you just identified. Because of its large size and exposed position, a wound in the superior part of this vein, where it is largest, is particularly dangerous because of extreme internal bleeding and also because of the risk of air entering the vessel, producing a blockage [embolus] in the blood returning to the heart and/or lungs.)

3. What vein is formed by the union of the subclavian vein with the internal jugular vein?

_____

4. What is the name of the large blue vein passing superiorly on the medial (ulnar) side of the arm?

_____

5. With what vein does the external jugular vein merge to return blood to the heart?

_____

**Beyond
A I A**

6. What is the function of the internal jugular vein?

_____

7. Through what structure does the internal jugular vein exit the skull?

_____

LYMPHATIC SYSTEM

STUDENT OBJECTIVES

OVERVIEW
- Review the anatomy of the lymphatic system.

LYMPHOID ORGANS
- Describe the location of various groups of lymph nodes within the body, including the cervical, axillary, and inguinal nodes.
- Describe the location of the cysterna chyli.
- Describe the location and structure of the thoracic duct.

- Describe the location and structure of special lymphoid organs, including the spleen and thymus.

LYMPH DRAINAGE
- Describe some specific lymphatic drainage patterns and associated structures for the head, neck, face, and tongue.

LYMPHOID ORGANS

Exercise 7.1 Lymph Nodes and Lymphatic Vessels

➤ *Open AIA by double-clicking Start Interactive Anatomy. Choose the Dissectible Anatomy button.*

➤ *Be sure that Male and the Anterior thumbnail icon are selected. Click Open.*

➤ *Expand the window and center the image on the chest.*

➤ *Adjust the Layer Indicator to 263, click the Highlight button in the Tool palette, and select Lymphatic from the System drop-down menu.*

➤ *In this image, the heart and its great vessels, as well as all remaining thoracic and abdominal viscera, have been removed.*

1. Use the Identify tool to label the structures in the following diagram.

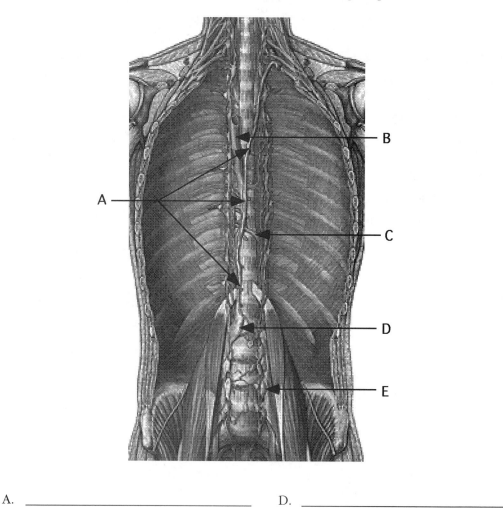

A. _____   D. _____

B. _____   E. _____

C. _____

2. What is the name of the expanded inferior end of the thoracic duct lying in front of the T12 vertebra and behind the aorta? (Remember: The aorta has been removed from the current image on your screen.)

_____

Beyond A I A   3. What does the name *azygous* mean, literally?

_____

4. The azygous vein joins the two largest veins in the body, which drain the superior and inferior halves of the body, respectively. Name these two veins.

_____

5. What is the function of the thoracic duct?

_____

6. What is the derivation of *cysterna chyli*?

_____

7. What are the two sources of extracellular fluid?

_____

8. What are the three functions of the lymphatic system?

   a. _____

   b. _____

   c. _____

9. What is lymph?

_____

10. What is the name of the pressure that forces fluids from the blood capillaries into the extracellular spaces?

_____

11. What are the two ways in which lymph vessels are similar to veins?

   a. _____

   b. _____

12. From what region(s) does the right lymphatic duct drain lymph?

_____

13. Into what specific blood vessel does the right lymphatic duct return lymph to blood circulation?

_____

14. From what region(s) does the thoracic duct (left lymphatic duct) drain lymph?

_____

15. Into what specific blood vessel does the thoracic duct return lymph to blood circulation?

_____

16. What is the name of the expanded sac in the abdomen where the thoracic duct originates?

_____

17. What type of cell is produced in the germinal centers in the medulla of a lymph node?

_____

18. What are the names of the two cells that provide important functions in the immune system?

_____

19. What is the largest lymphoid organ?

_____

20. What are the two types of tissues found with the spleen?

_____

21. What are the two types of cells found with the red pulp?

_____

22. What is the important role of the thymus?

_____

23. Specifically, what type of lymphocyte matures and develops with the thymus?

_____

24. What are the names of the three pairs of tonsils?

_____

➤ _Move the Navigator Box over the genital area of the image, zoom in, and adjust the Layer Indicator to 3._

25. Label the following diagram.

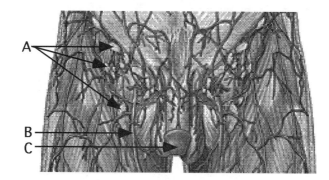

A. _____

B. _____

C. _____

(Note: Cancer of the scrotum [the fibromuscular sac that holds the testes] will metastasize [spread] to the superficial inguinal lymph nodes.)

Exercise 7.2 Spleen

➤ *Adjust the image so that the central abdominal region is visible, and set the Layer Indicator to 214.*

1. Zoom out and use the Identify tool to label the diagram below.

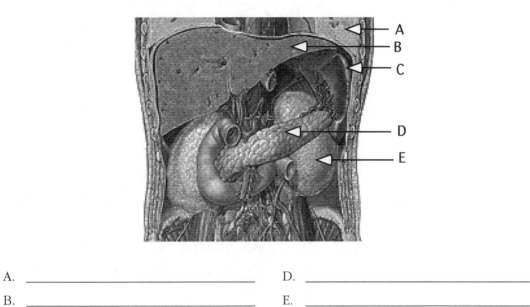

A. _____   D. _____

B. _____   E. _____

C. _____

2. What organ (not seen in this image) leaves an impression on the surface of the spleen?

   _____

3. Is the head or tail of the pancreas closest to the spleen?

   _____

4. What is the name of the large respiratory muscle that borders the superior and posterior edges of the spleen?

   _____

Beyond A I A 5. What is the function of the spleen?

   _____

Exercise 7.3 Thymus

➤ *Adjust the image so that the upper chest region is visible, and set the Layer Indicator to 164.*

1. Zoom in and then label the following diagram.

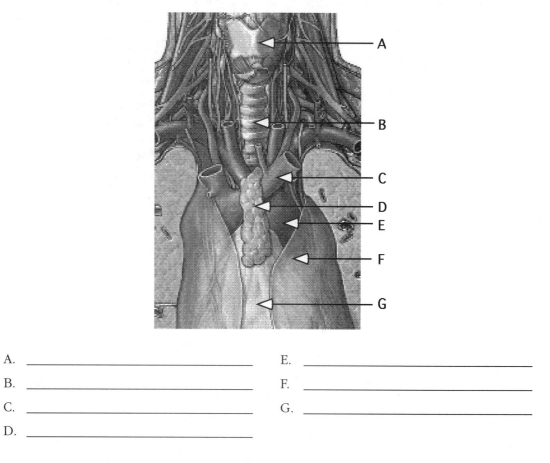

A. _____   E. _____

B. _____   F. _____

C. _____   G. _____

D. _____

2. Set the Layer Indicator to 154, and use the Identify tool to label the diagram below.

A. _____   D. _____

B. _____   E. _____

C. _____

(Note: The thymus is prominent in infants and diminishes in size, or atrophies, as one ages.)

LYMPH DRAINAGE

Exercise 7.4 Specific Lymphatic Drainage

➤ *Close the Male Anterior window. Select Open Content from the File menu.*

➤ *Click Atlas Anatomy.*

➤ *From the drop-down menus that appear, make the following selections:*

Body Region: Head and Neck   View Orientation: Lateral
Body System: Lymphatic   Image Type: Illustration

➤ *Click on the "Lymph Flow of Head" thumbnail icon. Click Open.*

➤ *Select Lymphatic from the System drop-down menu palette.*

1. Label the lettered pins on the diagram that follows.

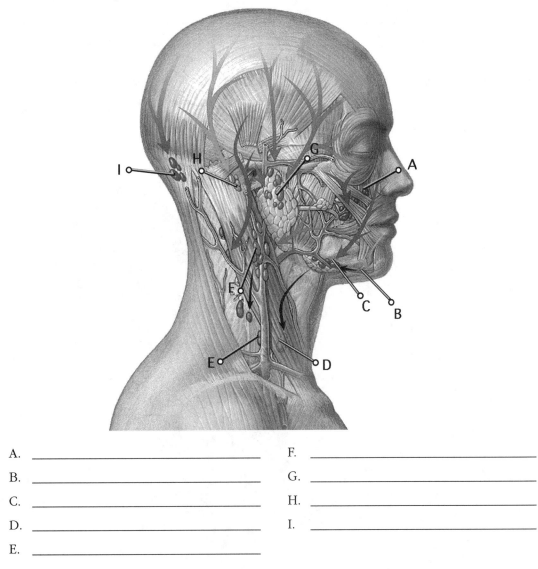

A. _____   F. _____

B. _____   G. _____

C. _____   H. _____

D. _____   I. _____

E. _____

2. Both the deep and superficial cervical lymph nodes lie on or close to what major strap muscle?

_____

3. What specific region does the buccal lymph node drain?

   _____

4. What is the function of the parotid duct?

   _____

5. What is the function of the submandibular lymph node?

   _____

6. What is the derivation of *submental lymph node?*

   _____

7. What is the function of the external jugular vein?

   _____

8. What does the term *retroauricular lymph node* imply about its anatomical position?

   _____

➤ *Close the "Lymph Flow of the Head (Lat)" window. Select Open Content from the File menu. From the View Orientation drop-down menu, select Superior.*

➤ *Select the "Lymph Flow of the Tongue (Dorsal)" thumbnail icon. Click Open.*

➤ *Expand the window, zoom out, and select Show All Pins from the Tool palette.*

9. Label the following diagram.

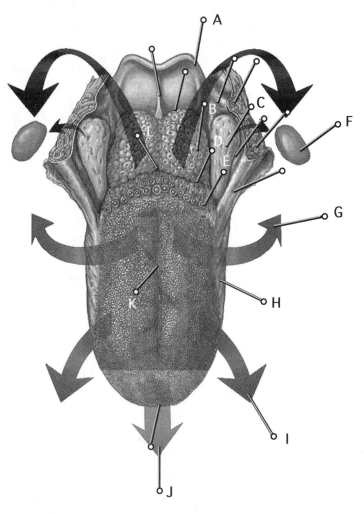

A. _____   G. _____

B. _____   H. _____

C. _____   I. _____

D. _____   J. _____

E. _____   K. _____

F. _____   L. _____

10. What is the name of the V-shaped groove that divides the dorsum of the tongue into an anterior and posterior part?

_____

11. What is the name of the small, median, pitlike depression located at the apex of the terminal sulcus?

_____

12. The collection of lymphoid nodules located behind the terminal sulcus is collectively known as what?

_____

Beyond A I A

13. What is the function of the epiglottis?

_____

14. What is the derivation of *foliate papillae*?

_____

15. What is the function of the submental lymph nodes?

_____

16. What is the function of the jugulodigastric lymph node?

_____

Clinical Animation

To observe a clinical animation of the specific lymphatic drainage of the breast, do the following:

➤ *Select Open Content from the File Menu. Select Clinical Animations. From the drop-down menus which appear, make the following selections:*

Body Region: Thorax
Body System: Lymphatic

➤ *Click on the "Lymphatics and the Breast" thumbnail icon. Click Open and expand the window by clicking the Maximize button.*

➤ *A "Lymphatics and the Breast" window will open and begin to play. Below the animation window, a text window can be scrolled through.*

➤ *After the animation finishes, close the "Lymphatics and the Breast" window.*

RESPIRATORY SYSTEM

STUDENT OBJECTIVES

OVERVIEW

- Review the anatomy of the respiratory system.
- Describe the pathway of air as it enters the mouth and nose, including the final exchange of oxygen for carbon dioxide that occurs at the alveolar-pulmonary capillary interface.

LUNGS

- Describe the gross anatomy of the lungs, including their lobes and fissures.
- Relate the structure of the diaphragm.
- Describe the formation and boundaries of the mediastinum.

LOWER AIRWAY

- Describe the structure of the lower airway, including the larynx, trachea, and principal bronchi.
- Using 3D Anatomy, identify various components of the respiratory system.

UPPER AIRWAY

- Identify the components of the upper airway, including the nose and pharynx.

LUNGS

Exercise 8.1 Gross Anatomy of the Lungs

➤ *Open AIA by double-clicking Start Interactive Anatomy, and select Dissectible Anatomy.*

➤ *Be sure that Male and the Anterior thumbnail icon are selected. Click Open.*

➤ *Expand the window and adjust it so that it is centered on the chest. Zoom in.*

➤ *Adjust the Layer Indicator to 162. Click the Highlight button in the Tool palette.*

1. Use the Identify tool to label the three lobes of the right lung and the two lobes of the left lung on the following diagram.

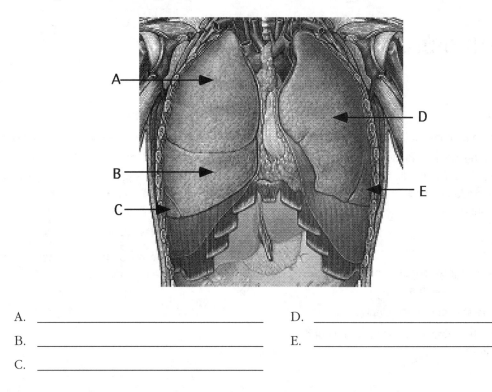

A. _____   D. _____

B. _____   E. _____

C. _____

2. The following diagram depicts the items found within the mediastinum and other structures associated with the superficial view of the lungs. Use the Identify tool to label the following diagram.

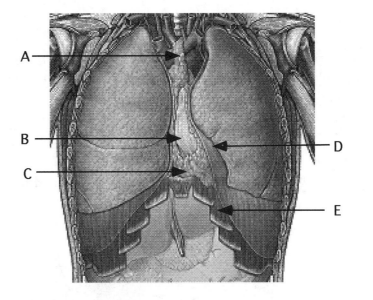

A. _____   D. _____

B. _____   E. _____

C. _____

Beyond
A I A

3. What is the mediastinum?

   _____

4. What is the significance of the lung region known as the cardiac notch?

   _____

5. What does *retrosternal fat pad* mean, literally?

   _____

6. What structure is contained within the pericardial sac?

   _____

7. What type of muscle tissue makes up the diaphragm?

   _____

8. What is the role of the diaphragm?

   _____

9. What is the function of the phrenic nerve?

   _____

10. Name the two fissures of the right lung that divide it into a superior, middle, and inferior lobe, respectively.

   _____

11. Name the fissure of the left lung.

   _____

Exercise 8.2 Left Lung

➤ *Close the Dissectible Anatomy window, and click on File Open Content. Click on the Atlas Anatomy button. From the drop-down menus that appear, make the following selections:*

Body Region: Thorax
Body System: Respiratory
View Orientation: Medial

➤ *Click on the "Left Lung (Med)" thumbnail icon. Click Open.*

1. Expand the window and identify the lettered pins in the diagram below.

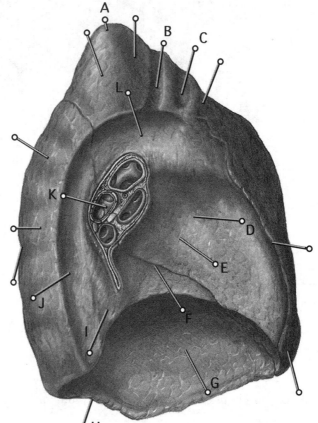

A. _____	G. _____
B. _____	H. _____
C. _____	I. _____
D. _____	J. _____
E. _____	K. _____
F. _____	L. _____

**Beyond
A I A**

2. What is the function of the oblique fissure of the left lung?

_____

➤ *Consider the left inferior pulmonary vein.*

3. What is its function?

_____

**Beyond
A I A**

4. Why is it depicted in red?

_____

5. Which heart chamber does it enter?

_____

6. What is the function of bronchopulmonary lymph nodes?

_____

7. What is the function of the left main bronchus?

_____

➢ *Consider the left pulmonary artery.*

8. What is its function?

_____

9. Why is it depicted in blue?

_____

10. Which heart chamber does this vessel exit as it proceeds to the lung?

_____

LOWER AIRWAY

Exercise 8.3 Larnyx

➢ *Close the Atlas Anatomy window. Click on File Open Content.*

➢ *Click on the Dissectible Anatomy button and the Male button.*

➢ *Click on the Anterior thumbnail icon. Click Open. Expand the window and set the Layer Indicator to 252.*

➢ *Adjust the image to match the following diagram and zoom in.*

1. Use the Identify tool to label the diagram below.

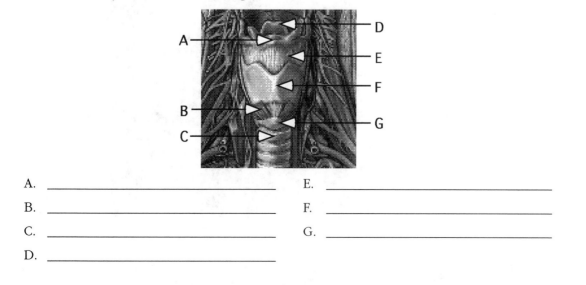

A. _____   E. _____

B. _____   F. _____

C. _____   G. _____

D. _____

2. What does the thyrohyoid membrane's name indicate about its location?

_____

3. What does the cricothyroid muscle's name indicate about its location?

_____

4. List the names of the three unpaired cartilages forming the larynx.

_____

_____

_____

5. Of the three cartilages listed in the answer to question 4, which is the largest?

_____

6. Of what specific type of cartilage do the tracheal rings consist?

_____

7. What is the function of the epiglottis?

_____

Exercise 8.4 Trachea and its Branches

➢ *Move the Layer Indicator to 256, and adjust the image to match the diagram below.*

1. Use the Identify tool to label the diagram.

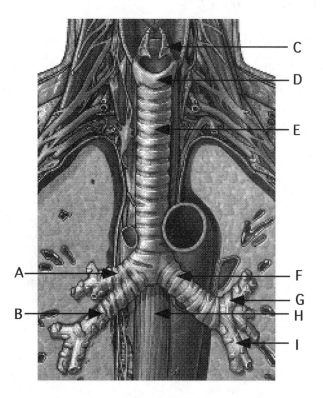

A. _____   F. _____

B. _____   G. _____

C. _____   H. _____

D. _____   I. _____

E. _____

2. What are the principal (primary, mainstem) bronchi?

    _____

3. Where does the trachea lie in relation to the esophagus?

    _____

**Beyond
A I A**
4. What is the role of the trachea?

    _____

5. What is the role of the esophagus?

    _____

6. What does *lobar bronchus* mean, literally?

    _____

7. Which principal bronchus, left or right, is more vertical?

    _____

8. Based on your answer above, which principal bronchus (left or right) would be more
    likely to be the site of lodged food? Why?

    _____

    _____

Exercise 8.5 Lower Airway Structures

➤ *Close the Dissectible Anatomy window. Select Open Content from the File menu.*

➤ *Click 3D Anatomy and select 3D Lungs. Click Open. An anterior 3D view of the lower
  airway and lungs will appear. Maximize the window.*

➤ *Click on the Structure List. Select Apex of the left lung from the list.*

➤ *Select the following terms from the Structure List, and then answer the questions that
  follow.*

1. Arytenoid cartilage
    Is this a paired or unpaired cartilage?

    _____

2. Carina
    a. Where is the carina located?

        _____

    b. What two tunnel-like openings lie on either side of the carina in this view?

        _____

3. Corniculate cartilage

   To what paired cartilages do these cartilages attach?

   _____

4. Cricoid cartilage

   Below which laryngeal cartilage does the cricoid cartilage lie?

   _____

5. Diaphragmatic surface(s) of the inferior lobe of the left and right lung

   Are these surfaces concave or convex? Why?

   _____

6. Epiglottis

   Describe the overall shape of this cartilage.

   _____

7. Hilum of the left lung

   In this view, is this region concave or convex?

   _____

**Beyond
A I A**

8. What important vessels enter or exit the lung at its hilum?

   a. Arteries: _____

   b. Veins: _____

   c. Air passages: _____

9. Horizontal fissure of the right lung

   How many lobes make up the right lung?

   _____

10. Hyoid bone

    a. Above which large laryngeal cartilage is the hyoid bone located?

    _____

    b. What is the name of the cartilage projecting just posterior to the hyoid bone in this view?

    _____

11. Vocal ligament

    In this view, what two laryngeal cartilages do these ligaments span?

    _____

**Beyond
A I A**

12. What is the derivation of *apex of the left lung*?

    _____

13. What is the function of the vocal ligament?

    _____

14. What is the derivation of *epiglottis*?

    _____

Exercise 8.6 Bronchial Tree

➤ *Close the 3D Lung window and select Open Content from the File menu.*

➤ *Click on the Atlas Anatomy button and make the following selections from the drop-down menus:*

Body Region: Thorax
Body System: Respiratory
View Orientation: Anterior
Image Type: Illustration

➤ *Click on the "Bronchial Tree" thumbnail icon. Click Open.*

1. Expand the window and label the following diagram.

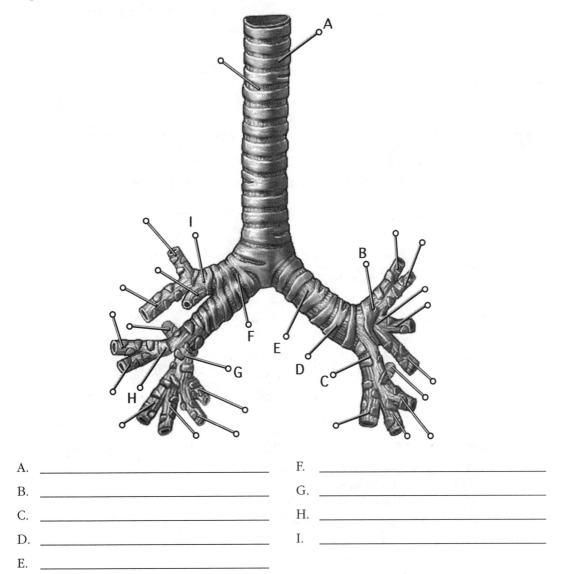

A. _____   F. _____

B. _____   G. _____

C. _____   H. _____

D. _____   I. _____

E. _____

2. How many lobar (secondary) bronchi supply the left lung?

_____

3. How many lobar (secondary) bronchi supply the right lung?

_____

Beyond A I A

4. What is the function of the tracheal rings?

_____

➤ *Compare the anterior and posterior views of the bronchial tree. Do not close the "Bronchial Tree (Ant)" window. Select Open Content from the File menu.*

➤ *From the View Orientation drop-down menu, select Posterior. Click on the "Bronchial Tree (P)" thumbnail icon. Click Open.*

➤ *Now select Tile Vertically from the Window menu. Zoom out in both the "Bronchial Tree (Ant)" and "Bronchial Tree (Post)" windows. The two windows now appear side by side on the screen.*

5. What noticeable difference in the structure of the tracheal rings exists between the anterior and posterior views?

_____

Beyond A I A

6. Explain the purpose of the open posterior portions of the tracheal rings.

_____

7. Is the entire lower airway supported by tracheal rings? Why?

_____

UPPER AIRWAY

Exercise 8.7 Nasal Conchae

➤ *Close the "Bronchial Tree" windows and click on the Dissectible Anatomy button. Click on the Medial view thumbnail icon. Click Open.*

➤ *Expand the window and center on the head. Set the Layer Indicator to 35.*

➤ *Select Open Content from the File menu. Click on the Atlas Anatomy button.*

➤ *From the drop-down menus that appear, make the following selections:*

Body Region: Head and Neck
Body System: Respiratory
View Orientation: Lateral
Image Type: Cadaver Photograph

➤ *Click on the "Lateral Wall of Palate" thumbnail icon. Click Open. Expand the window and zoom out on the image.*

➤ *Select Tile Vertically from the Window menu.*

1. Identify the structures in the following diagrams.

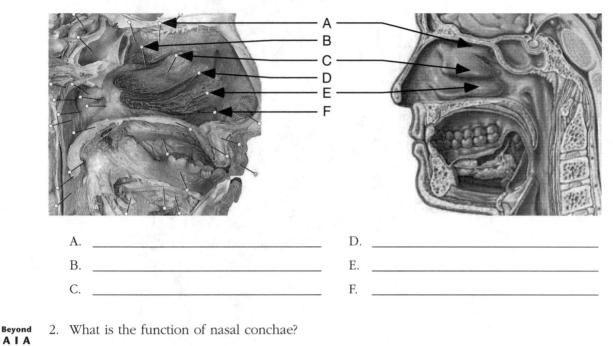

A. _____   D. _____

B. _____   E. _____

C. _____   F. _____

Beyond
A I A

2. What is the function of nasal conchae?

   _____

3. Why did early anatomists choose the term *conchae* to describe these structures?

   _____

   (Note: Another name for these structures is turbinates.)

Exercise 8.8 Pharynx

➤ *Close the "Lateral Wall of the Palate" window.*

➤ *Expand the Male Medial window.*

1. Use the Identify tool to label the following diagram with the terms below:

   - Esophagus
   - Nasopharynx
   - Oropharynx
   - Laryngopharynx
   - Epiglottis

   - Mucosa of trachea
   - Vestibular fold
   - Vocal fold
   - Pharyngeal orifice of auditory tube

A. _____   F. _____

B. _____   G. _____

C. _____   H. _____

D. _____   I. _____

E. _____

2. Into what portion of the pharynx does the mouth cavity empty?

    _____

3. In what pharyngeal region do the esophagus and trachea diverge?

    _____

4. What two structures does the auditory tube connect?

    _____

    _____

DIGESTIVE SYSTEM

STUDENT OBJECTIVES

OVERVIEW

- Review the anatomy of the digestive (alimentary) system.

MOUTH AND ESOPHAGUS

- Describe and compare the location of the three pairs of major salivary glands and their associated ducts.
- Describe the structure of the tongue.
- Identify the innervation of the tongue.
- Describe the types of teeth, including incisors, canines, premolars, and molars.
- Describe the dental formula for humans.

STOMACH AND ABDOMINAL CAVITY

- Describe the structure and location of the esophagus, stomach, small intestine, and large intestine.
- Identify the regions of the stomach, small intestine, and large intestine.
- Describe the structure and location of the liver, gall-bladder, and pancreas.
- Describe the special duct system of the liver, gallbladder, and pancreas.

MOUTH AND ESOPHAGUS

Exercise 9.1 Salivary Glands

➤ *Open AIA by double-clicking Start Interactive Anatomy and select Dissectible Anatomy.*

➤ *Be sure that the Male and the Lateral thumbnail icons are selected. Click Open. Expand the window.*

➤ *Maximize and adjust the primary window so that it is centered on the side of the face.*

➤ *Adjust the Layer Indicator to 10. Click the Highlight button in the Tool palette.*

1. Label the diagram below using the Identify tool.

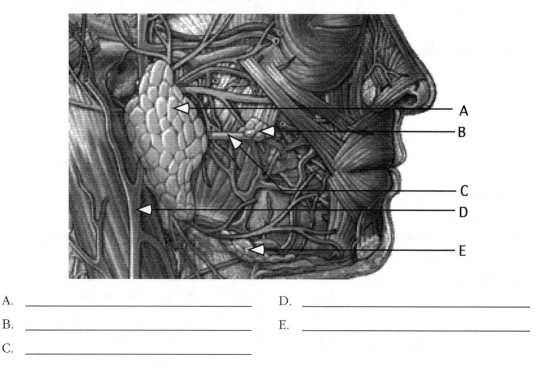

A. _____   D. _____

B. _____   E. _____

C. _____

2. Adjust the Layer Indicator to 227, and then label the diagram below.

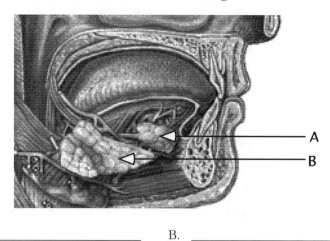

A. _____   B. _____

3. What is the derivation of the term *sublingual*?

_____

4. What is the derivation of the term *submandibular*?

_____

5. What is a bolus?

_____

6. What is the role of the enzyme salivary amylase?

_____

7. Saliva is primarily responsible for the digestion of what major food type?

_____

Exercise 9.2   Tongue

1. Adjust the Layer Indicator to 242, and then label the diagram below.

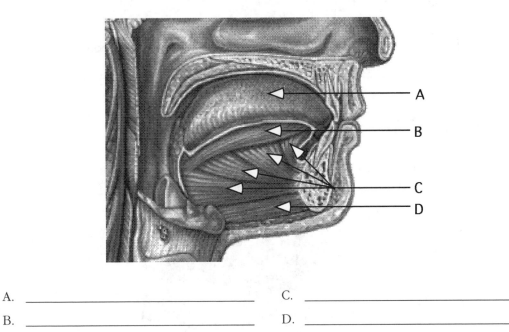

A. _____   C. _____

B. _____   D. _____

2. Adjust the Layer Indicator to 233, and label the diagram below.

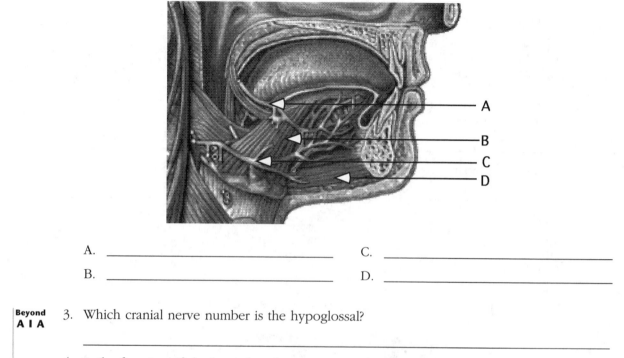

A. _____   C. _____

B. _____   D. _____

Beyond
A I A 3. Which cranial nerve number is the hypoglossal?

_____

4. Is the function of the hypoglossal nerve sensory, motor, or both?

_____

Exercise 9.3 Teeth

1. Set the Layer Indicator to 162, and label the diagram below.

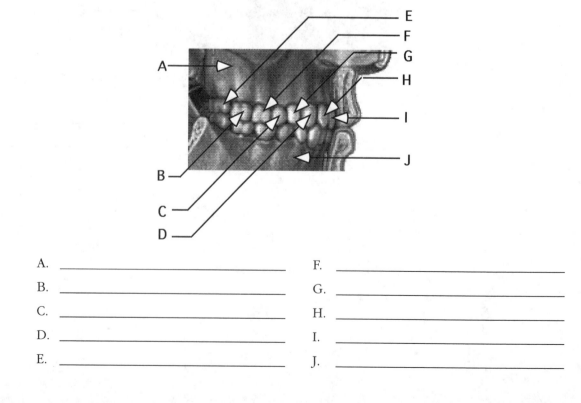

A. _____   F. _____

B. _____   G. _____

C. _____   H. _____

D. _____   I. _____

E. _____   J. _____

2. Describe the function of incisors.

_____

3. Describe the function of canines.

_____

4. Describe the function of premolars.

_____

5. Describe the function of molars.

_____

6. The dental formula for humans is $\frac{2\text{-}1\text{-}2\text{-}3}{2\text{-}1\text{-}2\text{-}3}$. What does this formula mean?

_____

Exercise 9.4 Esophagus

➤ *Adjust the image so that the right lateral chest region occupies the screen. Set the Layer Indicator to 223.*

1. Adjust the image as needed to label the diagram.

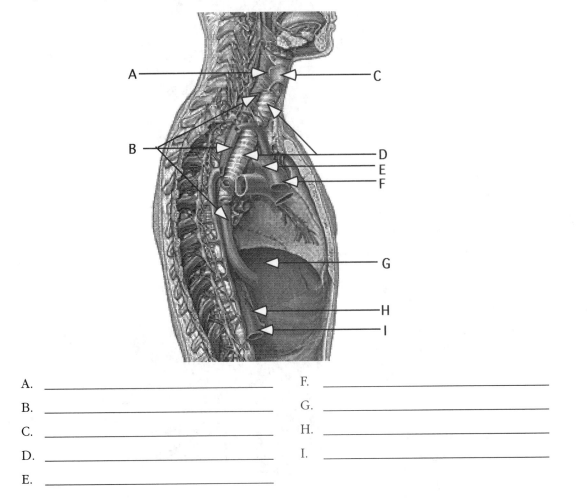

A. _____   F. _____

B. _____   G. _____

C. _____   H. _____

D. _____   I. _____

E. _____

2. The esophagus is located _____ to the trachea.

3. What is peristalsis?

_____

4. What is the role of the epiglottis?

_____

5. What is the esophageal hiatus?

_____

STOMACH AND ABDOMINAL CAVITY

Exercise 9.5 Stomach

➤ *Select Anterior from the View button drop-down menu.*

➤ *Set the Layer Indicator to 204, and adjust the image to match the diagram below.*

1. Use the Identify tool to label the following diagram:

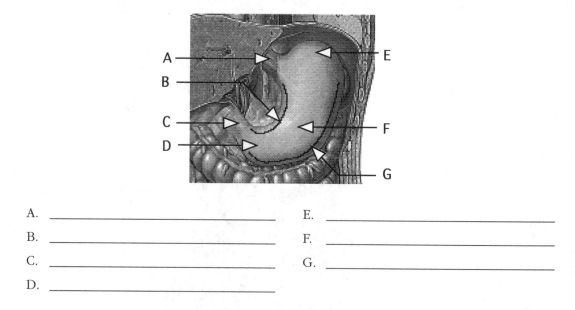

A. _____   E. _____

B. _____   F. _____

C. _____   G. _____

D. _____

2. Set the Layer Indicator to 205, and label the diagram below.

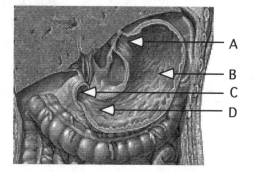

A. _____   C. _____

B. _____   D. _____

➤ *Select Lateral from the View button drop-down menu.*

➤ *Center the image over the stomach region.*

➤ *Set the Layer Indicator to 173.*

3. Label the following diagram.

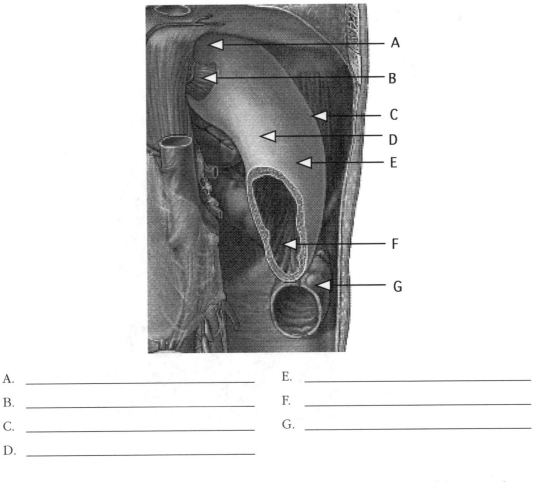

A. _____   E. _____

B. _____   F. _____

C. _____   G. _____

D. _____

➤ *Answer the following questions, which pertain to the previous diagrams of the stomach.*

4. What two structures in the gastrointestinal tract are connected at the cardiac orifice? (Note: Oral refers to closest to the mouth; aboral refers to farthest from the mouth.)

a. Orally: _____

b. Aborally: _____

5. What two structures in the gastrointestinal tract are connected by the pyloric sphincter?

a. Orally: _____

b. Aborally: _____

Beyond
A I A
6. What are rugae?

_____

7. What is the cardiac portion of the stomach?

_____

8. Describe a hiatal hernia in terms of the stomach, esophagus, and esophageal hiatus.

_____

Exercise 9.6 Abdominal Cavity

➤ *Select Anterior from the View button drop-down menu. Set the Layer Indicator to 195, adjust the image to match the diagram below.*

1. Label the diagram.

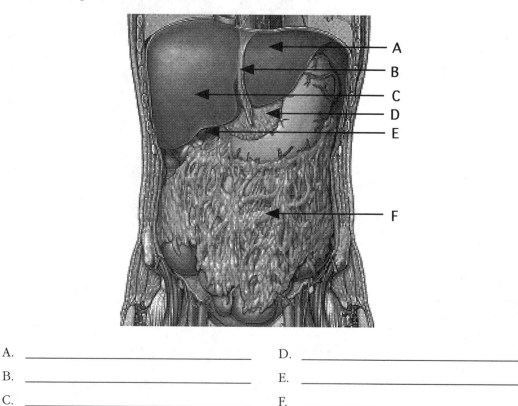

A. _____   D. _____

B. _____   E. _____

C. _____   F. _____

2. The lesser omentum arises from what part of the stomach?

_____

3. The greater omentum arises from what part of the stomach?

_____

Beyond
A I A
4. What is the function of the omenta?

_____

Exercise 9.7 Small and Large Intestines

➤ *Set the Layer Indicator to 208. (Note: In this view, the transverse colon of the large intestine and the stomach have been removed, revealing the pancreas covered with peritoneum.)*

1. Label the following diagram.

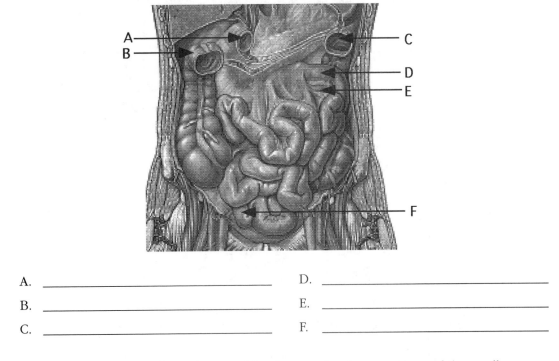

A. _____   D. _____

B. _____   E. _____

C. _____   F. _____

2. Moving aborally (away from the mouth), what are the three portions of the small intestine?

    a. Proximal: _____

    b. Middle: _____

    c. Distal: _____

Beyond A I A

3. What is the function of mesentery?

    _____

➤ *To see the attachment of the small intestine to the large intestine, set the Layer Indicator to 209. In this view, the jejunum and most of the ileum have been removed, and the cut ends of the duodenum can be seen lying behind the peritoneal layer.*

4. Use the Identify tool to label the following diagram.

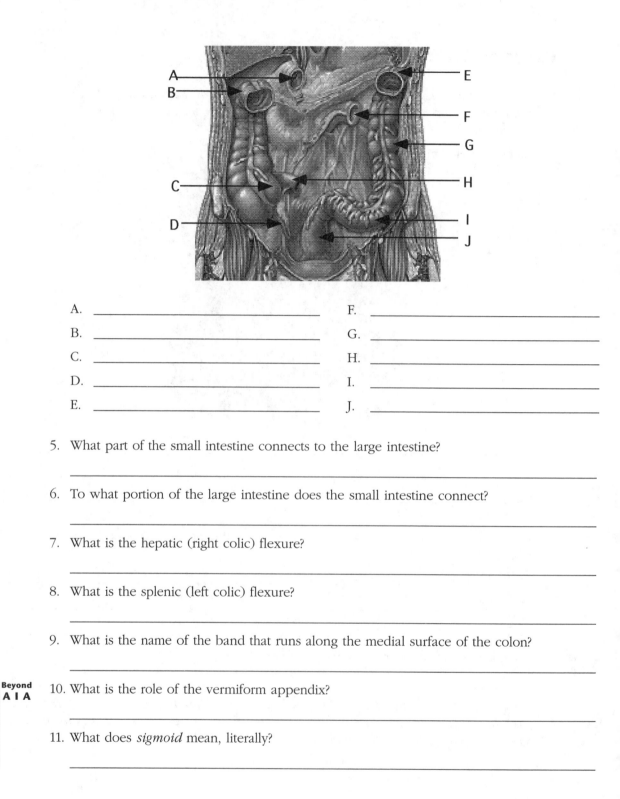

A. _____   F. _____

B. _____   G. _____

C. _____   H. _____

D. _____   I. _____

E. _____   J. _____

5. What part of the small intestine connects to the large intestine?

   _____

6. To what portion of the large intestine does the small intestine connect?

   _____

7. What is the hepatic (right colic) flexure?

   _____

8. What is the splenic (left colic) flexure?

   _____

9. What is the name of the band that runs along the medial surface of the colon?

   _____

**Beyond
A I A**   10. What is the role of the vermiform appendix?

   _____

11. What does *sigmoid* mean, literally?

   _____

Exercise 9.8 Liver, Gallbladder, and Associated Duct Systems

1. Set the Layer Indicator to 198, and label the following diagram.

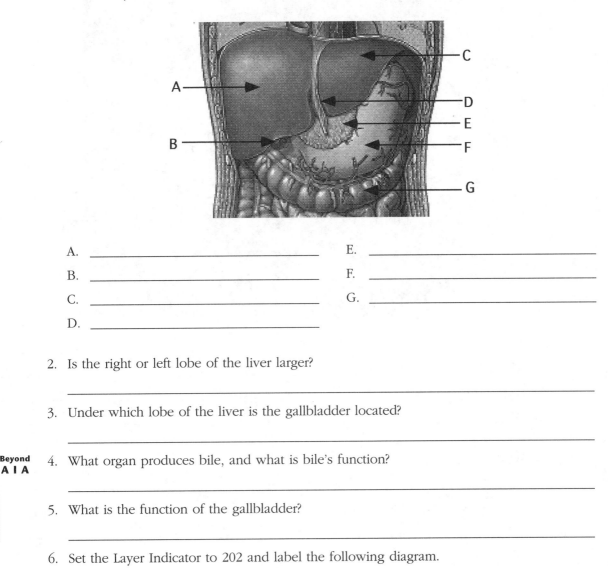

A. _____   E. _____

B. _____   F. _____

C. _____   G. _____

D. _____

2. Is the right or left lobe of the liver larger?

    _____

3. Under which lobe of the liver is the gallbladder located?

    _____

Beyond
A I A

4. What organ produces bile, and what is bile's function?

    _____

5. What is the function of the gallbladder?

    _____

6. Set the Layer Indicator to 202 and label the following diagram.

A. _____   E. _____

B. _____   F. _____

C. _____   G. _____

D. _____   H. _____

7. What is the cystic duct?

_____

8. What is the hepatic duct?

_____

**Beyond
A I A**
9. What is the function of the portal vein?

_____

10. The portal vein, in fulfilling its function, supplies the liver with most of its nutritional needs. What is the function of the hepatic artery?

_____

11. Set the Layer Indicator to 214. In this image, four regions of the pancreas are identified from its distal (free) to its proximal (attached) end. In order from the distal end, what are they? (Note: Use the Highlight feature to help identify these regions.)

a. (Distal) _____

b. _____

c. _____

d. (Proximal) _____

12. Adjust the Layer Indicator to 215. This removes the anterior half of the pancreas so that you can see the duct system of the pancreas and surrounding structures. Label the following diagram.

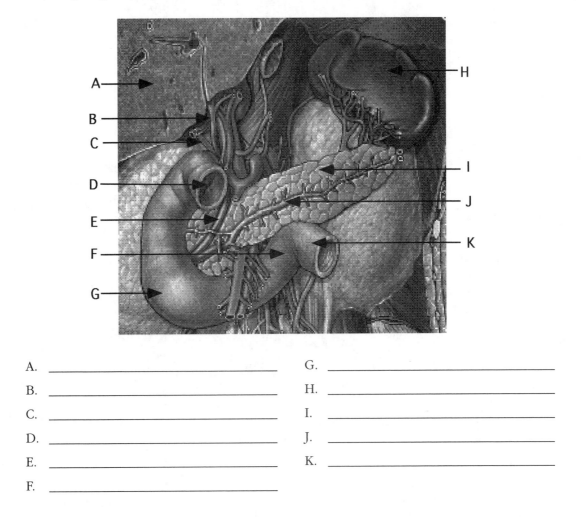

A. _____   G. _____

B. _____   H. _____

C. _____   I. _____

D. _____   J. _____

E. _____   K. _____

F. _____

13. What duct is formed by the merger of the cystic duct and the common hepatic duct?

_____

14. Where does the common bile duct empty its contents?

_____

15. From what branch point of the abdominal aorta do the common hepatic artery, left gastric artery, and splenic artery arise?

_____

Beyond A I A 16. What is the function of the spleen?

_____

URINARY SYSTEM

STUDENT OBJECTIVES

OVERVIEW

- Review the anatomy of the urinary system.
- Trace the production and modification of urine from the renal corpuscle to the urethral orifice.

KIDNEYS

- Describe the renal artery and its branches within the kidney.
- Describe the renal vein and its tributaries within the kidney.
- Identify and describe the functions of the various connective tissues surrounding the kidneys, including the renal fascia and renal capsule.
- Describe the location of the adrenal glands, each kidney, and the ureter.
- Identify the major urine-producing regions of the kidney, the cortex, and medulla.
- Describe the location of the renal pyramids and renal columns.
- Describe the location of the renal sinus.

URINE PATHWAY AND URINARY BLADDER

- Trace the production and elimination route of urine from the renal pyramids through the renal pelvis, ureters, urinary bladder, and urethra.
- Describe the structure, function, and location of the urinary bladder.

INTERNAL STRUCTURES OF THE KIDNEY

- Describe the structure and function of the functional unit of the kidney and the uriniferous tubule, consisting of the nephron and collecting duct.
- Identify the components of the juxtaglomerular apparatus, and describe the important role of this structure in regulating blood pressure.

KIDNEYS

Exercise 10.1   Arterial Supply to and Venous Return from the Kidneys

➤ *Open AIA by double-clicking Start Interactive Anatomy, and select Dissectible Anatomy.*

➤ *Select Male and Anterior. Click Open. Expand the window.*

➤ *Maximize and adjust the window so that it is centered on the abdomen.*

➤ *Adjust the Layer Indicator to 238.*

1. What is the name of the large blue vessel that runs vertically between the two kidneys?

_____

2. What are the names of the branches of this vessel that lead from the kidneys?

_____

Beyond AIA

3. Into what chamber of the heart does the inferior vena cava empty its contents?

_____

➤ *Set the Layer Indicator to 240. In this view, the abdominal aorta appears as a large red vessel between the kidneys.*

4. Use the Identify tool to label the following diagram of the branches of the aorta.

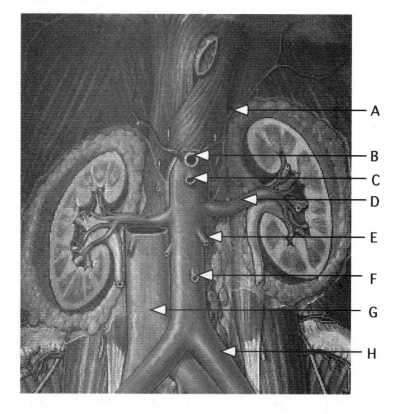

A. _____   E. _____

B. _____   F. _____

C. _____   G. _____

D. _____   H. _____

5. What blood vessels branch from the abdominal aorta (in this view) and lead to the kidney?

_____

6. Where the inferior phrenic arteries emerge from the abdominal aorta, there appears to be a hole. This is actually another branch point of the abdominal aorta. It is the:

_____

7. Directly inferior to this hole (answer to 6) is the exit point of another artery. This is the:

_____

8. What is the name of the branches of the abdominal aorta that lead to the kidneys?

_____

9. Below the branch points to the kidneys, two arteries arise that leave the lateral sides of the aorta (the left branch is slightly superior to the right). These are the left and right:

_____

10. Continuing inferiorly, another vessel arises from the medial portion of the abdominal aorta. This is the:

_____

11. At the inferior end of the abdominal aorta, it splits into the left and right:

_____

Exercise 10.2 Surface Features of the Kidney and Surrounding Structures

➤ *Adjust the Layer Indicator to 228.*

1. What is the name of the whitish-yellow material surrounding each of the kidneys in this view?

_____

Beyond A I A

2. What type of connective tissue is renal fascia?

_____

➤ *Adjust the Layer Indicator to 232. Click the Highlight button, and identify the left and right suprarenal glands.*

3. What is the anatomical location of the suprarenal glands?

_____

4. What is the common name for the suprarenal glands?

_____

➤ *Restore the image to full color by clicking the Normal button.*

5. Which of the two kidneys is at a superior position?

_____

6. What is the name of the yellow material at the edge of each kidney in this view?

_____

➤ *Locate the abdominal aorta within this view.*

7. Identify the lime-green structures that lie above and below the branch point of the inferior mesenteric artery from the abdominal aorta.

_____

_____

Beyond
A I A

8. To what body system do these lime-green structures belong?

_____

➤ *Click the yellow netlike structure covering the abdominal aorta.*

9. To what body system does this netlike structure belong?

_____

➤ *Adjust the Layer Indicator to 234.*

➤ *Select Open Content from the File menu. Choose Atlas Anatomy.*

➤ *Select Abdomen from the Region menu and Urinary from the System menu. Select the "Dissection of Kidneys (Ant)" thumbnail icon. Click Open.*

➤ *Select Tile Vertically from the Window menu.*

10. Adjust the images in both windows so that they match the following diagram. Use the Identify tool to label the diagram.

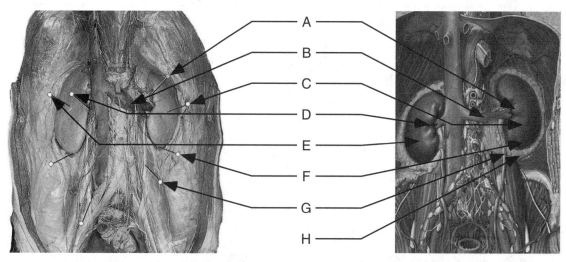

A. _____   E. _____

B. _____   F. _____

C. _____   G. _____

D. _____   H. _____

11. The ureter functions to carry _____ from the kidney

to the _____ .

12. What type of epithelial tissue lines the lumen of the ureter, and what is its function?

_____

Exercise 10.3 Internal Structures of the Kidney

➤ *Close the "Dissection of Kidneys (Ant)" window.*

➤ *Expand the "Dissectible Anatomy Male Anterior" window.*

➤ *Adjust the Layer Indicator to 237. The kidney has now been cut through the coronal plane.*

1. Identify the name given to the outer edge of the kidney (whitish-pink border).

_____

2. The middle portion of the kidney is subdivided into:

   a. darker pink regions called:

_____

   b. whitish-pink areas between the darker pink regions, called:

_____

3. Label the following diagram.

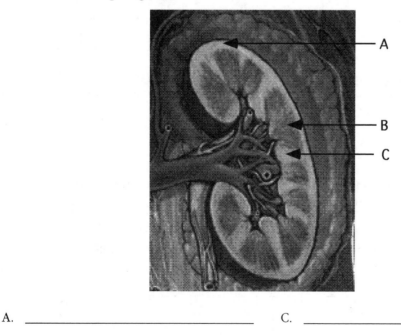

A. _____   C. _____

B. _____

➤ *Adjust the Layer Indicator to 243. The central medial region of the kidney appears with blood vessels and the renal pelvis and renal calyces extracted.*

4. Identify the large, hollow, concave region in the kidney in which these structures lie.

_____

URINE PATHWAY AND URINARY BLADDER

Exercise 10.4 Urine Pathway

➤ *Adjust the Layer Indicator to 242.*

1. Label the following diagram.

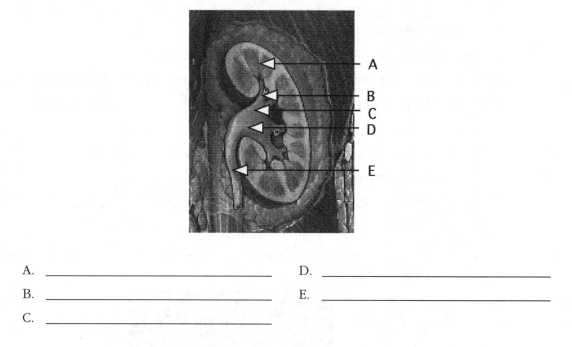

A. _____   D. _____

B. _____   E. _____

C. _____

➤ *Urine is formed in the renal pyramids of the kidney. The urine travels to the light tan structures found within the renal sinus. Follow the pathway from the lateral region of the kidney to the ureter.*

2. What is the name of the structures that directly collect urine from the renal pyramids?

_____

3. Urine from a group of these narrow regions (identified in question 2) is then pooled into larger areas directly medial to them. They are the:

_____

4. From the larger, medial regions (identified in question 3) urine is then collected in a structure that leads to the ureter. Identify this structure.

_____

Exercise 10.5 Medial View of the Urinary Bladder

➤ *Select Medial from the View button drop-down menu in the Tool palette. Set the Layer Indicator to 51.*

➤ *Identify the ureter and the urinary bladder.*

1. What is the name of the tube leading from the urinary bladder out of the body?

   _____

2. Identify the three portions of the urethra in a male as it passes through the following structures:

   a. Prostate gland: _____

   b. Sphincter urethra muscle: _____

   c. Penis: _____

INTERNAL STRUCTURES OF THE KIDNEY

Exercise 10.6 Internal Structures of the Kidney

➤ *Close the Male Medial window.*

➤ *Select Open Content from the File menu. Choose Atlas Anatomy.*

➤ *Select Abdomen from the Region menu and Urinary from the System menu. Select the "Renal Arteries" thumbnail icon. Click Open.*

➤ *Select Urinary from the Select System drop-down menu.*

➤ *Expand the window and label the diagram that follows.*

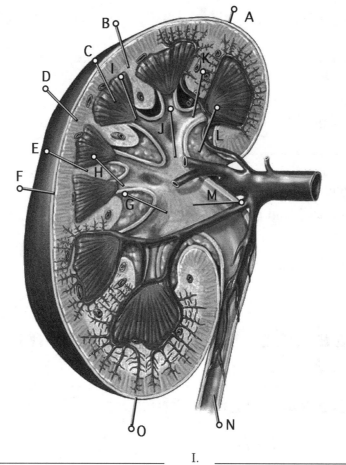

A. _____ I. _____

B. _____ J. _____

C. _____ K. _____

D. _____ L. _____

E. _____ M. _____

F. _____ N. _____

G. _____ O. _____

H. _____

Exercise 10.7 Diagram of a Nephron

➤ *Close the Renal Arteries window.*

➤ *Select Open Content from the File menu. Choose Atlas Anatomy.*

➤ *Select Abdomen from the Region menu and Urinary from the System menu. Select the "Diagram of Nephron" thumbnail icon. Click Open.*

➤ *Expand the window.*

1. Label the following diagram.

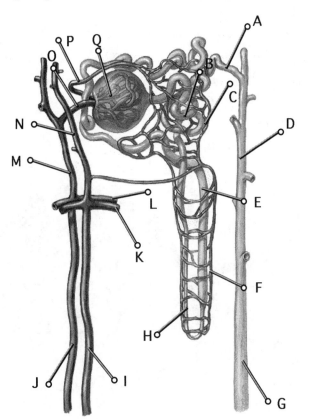

A. _____   J. _____

B. _____   K. _____

C. _____   L. _____

D. _____   M. _____

E. _____   N. _____

F. _____   O. _____

G. _____   P. _____

H. _____   Q. _____

I. _____

2. Which vessel carries blood into the glomerulus?

_____

3. What is the function of the arcuate vein?

_____

4. What is the function of the glomerulus?

_____

5. What is a nephron?

_____

Exercise 10.8 Diagram of Renal Glomerulus

➤ *Close the Diagram of Nephron window.*

➤ *Select Open Content from the File menu. Choose Atlas Anatomy.*

➤ *Select Abdomen from the Region menu and Urinary from the System menu. Select the "Diagram of Renal Glomerulus" thumbnail icon. Click Open.*

➤ *Expand the window.*

1. Label the diagram that follows.

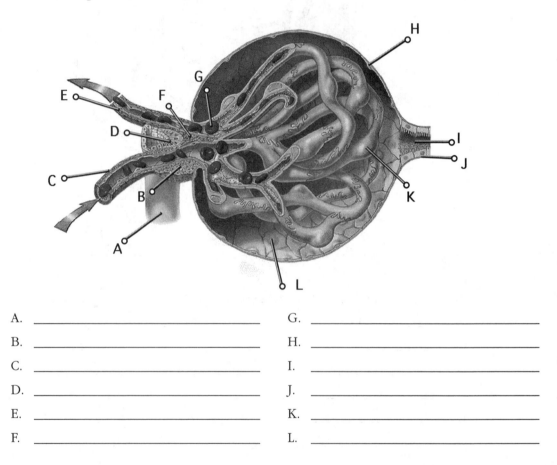

A. _____   G. _____

B. _____   H. _____

C. _____   I. _____

D. _____   J. _____

E. _____   K. _____

F. _____   L. _____

2. In this diagram, what is the name of the special group of deep purple cells (letter B) that line the afferent arteriole and make physical contact with the walls of the distal convoluted tubule?

_____

3. What name is given to the group of cells (letter D) of the walls of the distal convoluted tubule that are making contact with the juxtaglomerular cells of the afferent arteriole?

_____

4. From what specific type of tissue are juxtaglomerular cells derived?

_____

5. What important enzyme is released by juxtaglomerular cells?

_____

6. Are juxtaglomerular cells appropriately named? Explain.

_____

7. What is the literal meaning of *macula densa*?

_____

8. To what special cellular feature does macula densa refer?

_____

9. What two types of cells make up the juxtaglomerular apparatus?

_____

10. What is the special function of the juxtaglomerular apparatus?

_____

11. To which layer (visceral or parietal) of Bowman's capsule do the special cells known as podocytes belong?

_____

12. What is the function of podocytes?

_____

13. What is the function of red blood cells?

_____

14. What is the significance of the red blood cells shown moving single file through the afferent and efferent arteriole and the glomerulus?

_____

REPRODUCTIVE SYSTEMS

STUDENT OBJECTIVES

OVERVIEW

- Review the anatomy of the male and female reproductive (genital) systems.

MALE REPRODUCTIVE SYSTEM

- Describe the location and structure of the male external genitalia.
- Describe the blood supply to the penis, including the dorsal artery of the penis.
- Describe the location, structure, and function of the male internal genitalia.
- Identify the structures of the male external and internal genitalia in a midsagittal view.
- Describe the location and structure of the associated structures of the spermatic cord.
- Describe the location of the important bony landmarks of the male pelvis.

FEMALE REPRODUCTIVE SYSTEM

- Describe the location and structure of the female external genitalia.
- Describe the location and structure of the female internal genitalia.
- Identify the female genitalia (both external and internal) and associated structures in a midsagittal view.
- Identify the mammary gland and its associated structures.

MALE REPRODUCTIVE SYSTEM

Exercise 11.1 External and Internal Genitalia

➤ *Open AIA by double-clicking Start Interactive Anatomy, and choose Dissectible Anatomy.*

➤ *Select Male and Anterior. Click Open. Expand the window.*

➤ *Adjust the window so that it is centered on the male's external genitalia.*

➤ *Adjust the Layer Indicator to 18.*

1. What is the name of the long, central, white band that connects the superior end of the penis to the abdomen?

   _____

2. What is the name of the flesh-colored distal end (or head) of the penis?

   _____

➤ *Set the Layer Indicator to 21.*

3. What is the name of the midline vein on the dorsal side of the penis?

   _____

4. What is the name of the two arteries lying lateral to the midline vein?

   _____

➤ *Change the Layer Indicator to 27.*

5. What is the name of the long red muscle strands descending from the left and right inguinal areas of the abdomen down to and surrounding the testicles?

   _____

Beyond AIA 6. What is the function of the cremaster muscle?

   _____

➤ *Set the Layer Indicator to 156. Identify the long red arteries, the cremasteric arteries, as they pass up from the testes into the abdomen. Now trace the cremasteric artery superiorly to where it joins another artery. (Note: You will need to adjust the image to see where this artery emerges from a layer of overlying fascia.)*

7. What is the name of the artery that supplies the cremasteric artery?

   _____

8. What is the name of the overlying fascia covering the inferior epigastric artery?

   _____

➤ *Adjust the Layer Indicator to 179. Note the two long, paired arteries descending from each side of the lower abdomen onto the posterior aspect of each testis.*

9. Examining the right testicle (anatomical right), what is the name of the more lateral of the two arteries?

   _____

10. What is the name of the more medial of the two arteries?

   _____

11. Set the Layer Indicator to 180. Use the Identify tool to label the diagram below.

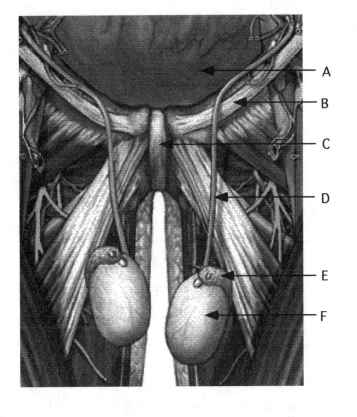

A. _____   D. _____

B. _____   E. _____

C. _____   F. _____

12. What is another name for the ductus deferens, and what does this other name imply about its function?

_____

(Note: Birth control in males can be achieved surgically by cutting a small section of the vas deferens in a simple outpatient procedure known as a *vasectomy*. The incision is made on the lateral side of the scrotal sac just superior to each testis, a small loop of the ductus deferens is pulled out, and a small section is excised. This interrupts the flow of sperm into the ejaculatory fluids and is thus a very reliable method of birth control.)

Clinical Animation

To observe a clinical animation of a vasectomy, do the following:

➤ *Select Open Content from the File Menu. Select Clinical Animations. Select Reproductive from the Body System drop-down menu. Select Vasectomy. Click the Open button. Expand the window by clicking the Maximize button.*

➤ *After the animation finishes, close the window.*

13. What structures are contained in the spermatic cord?

_____

➤ *Now trace the path of the ductus deferens as it travels superiorly from each testis, turning laterally and disappearing deep to the parietal peritoneum.*

14. What large vein does the ductus deferens cross as it passes deep to the parietal peritoneum above the inguinal ligament?

_____

15. Once the iliac vein passes below the inguinal ligament, what is it called?

_____

➤ *Select Medial from the View button drop-down menu. Set the Layer Indicator to 49. The ductus deferens is now visible superiorly, posteriorly, and inferiorly to the urinary bladder. Trace the ductus deferens and identify its parts as it runs from behind the bladder to the urethra. Two distinct regions can be identified in this midsagittal view along its course.*

16. What is the name of the region anterior to the seminal vesicle?

_____

17. What is the name of the region within the prostate gland?

_____

18. Label the diagram below.

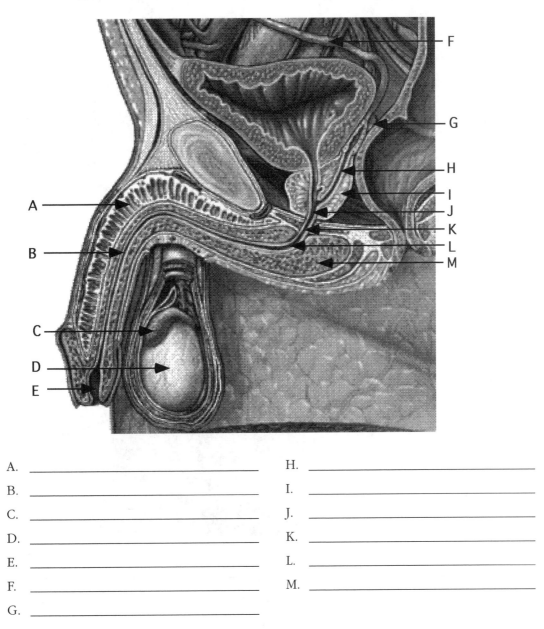

A. _____			H. _____	
B. _____			I. _____	
C. _____			J. _____	
D. _____			K. _____	
E. _____			L. _____	
F. _____			M. _____	
G. _____				

19. What are the names given to the three sections of the urethra?

_____

Beyond
A I A

20. What is the urogenital diaphragm?

_____

(Note: In addition to supporting the popular notion that the pelvic floor is formed almost exclusively by the levator ani and coccygeus muscles, many physicians and anatomists consider your answer to question 20 part of the pelvic floor because weaknesses that can occur during childbirth lead to herniation.)

Exercise 11.2 Body Wall of the Trunk (Anterior)

➤ *Close the Male Medial window.*

➤ *Select Open Content from the File menu. Choose Atlas Anatomy.*

➤ *Select Body Wall and Back from the Region menu and Anterior from the View menu. Select the "Body Wall of Trunk (Ant)" thumbnail icon. Click Open.*

➤ *Expand the window and label the diagram that follows.*

1. Adjust the image so that you can label the lettered pins on the following diagram. (Note: The figure below has been edited to illustrate relevant male structures.)

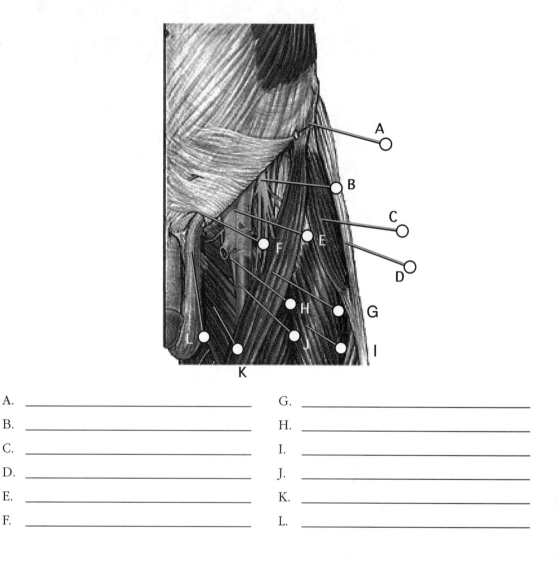

A. _____	G. _____
B. _____	H. _____
C. _____	I. _____
D. _____	J. _____
E. _____	K. _____
F. _____	L. _____

2. What special structure in the male emerges from the superficial inguinal ring?

_____

Beyond A I A

3. What is the femoral triangle?

_____

4. What muscle forms the lateral boundary of the femoral triangle?

_____

Exercise 11.3 Bony Landmarks of the Male Pelvis

➤ *Close the "Body Wall of Trunk (Ant)" window.*

➤ *Select Open Content from the File menu. Choose Atlas Anatomy.*

➤ *Select Pelvis and Perineum from the Region menu and Anterior from the View menu. Select the "Bony Landmarks of Male Pelvis" thumbnail icon. Click Open.*

➤ *Expand the window.*

1. Label the following diagram.

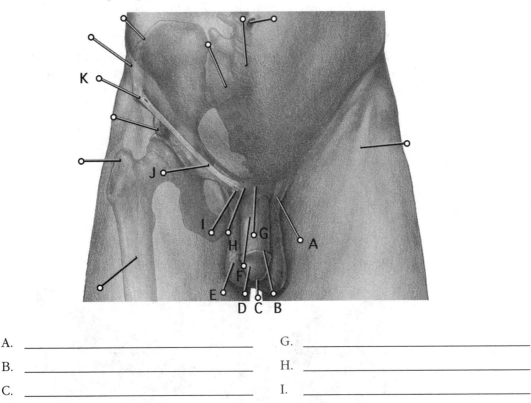

A. _____   G. _____

B. _____   H. _____

C. _____   I. _____

D. _____   J. _____

E. _____   K. _____

F. _____

2. What is the name of the strong fibrous band cord (letter J) running from the anterior superior iliac spine to the pubic tubercle?

_____

3. What is the name of the structure corresponding to the smooth distal end of the penis?

_____

4. What is the name of the structure that suspends the testes in the scrotum?

_____

5. What is the literal meaning of the word *corona*?

_____

6. What is another name for the external meatus of the urethra?

_____

Exercise 11.4 Male Superficial Perineal Space

➤ *Close the "Bony Landmarks of Male Pelvis" window.*

➤ *Select Open Content from the File menu. Choose Atlas Anatomy.*

➤ *Select Pelvis and Perineum from the Region menu, Reproductive from the System menu, and Inferior from the View menu. Select the "Male Super. Perineal Space 1" thumbnail icon. Click Open.*

➤ *Expand the window.*

➤ *The image that appears shows an inferior view of the male pelvis.*

1. Identify the lettered structures in the following diagram.

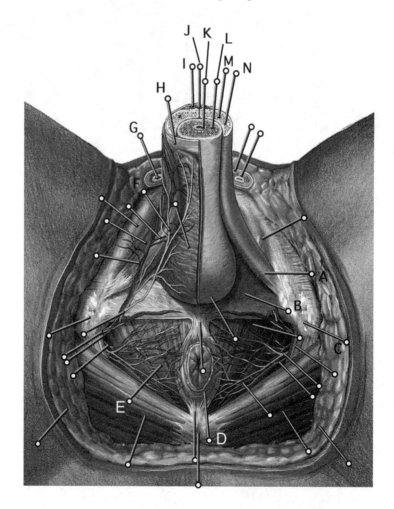

A. _____   H. _____

B. _____   I. _____

C. _____   J. _____

D. _____   K. _____

E. _____   L. _____

F. _____   M. _____

G. _____   N. _____

Beyond A I A

2. What is the function of the ductus deferens?

   _____

3. What two muscles contribute to the formation of the urogenital diaphragm?

   _____

4. What is the function of the external anal sphincter?

   _____

5. The spongy part of the urethra is identified in this view. What are the names of the two other portions of the urethra?

   _____

6 What is the function of the corpus cavernosum?

   _____

7. What is the literal meaning of the *crus* of the penis?

   _____

8. What is the principal function of the deep artery of the penis?

   _____

9. What is the principal function of the deep dorsal vein of the penis?

   _____

10. What is the name of the structure that supplies sensory fibers to both the skin of the penis and the glans penis?

   _____

Clinical Animation

To observe a clinical animation of sperm production and its pathway of ejaculation, do the following:

➢ *Select Open Content from the File Menu. Select Clinical Animations. Select Reproductive from the Body System drop-down menu. Select Sperm Production and Pathway of Ejaculation. Click the Open button. Expand the window by clicking the Maximize button.*

➢ *After the animation finishes, close the window.*

FEMALE REPRODUCTIVE SYSTEM

Exercise 11.5 External Genitalia

➤ *Close the Atlas Anatomy window and select Open Content from the File menu. Choose Dissectible Anatomy.*

➤ *Select Female and Anterior. Click Open.*

➤ *Expand the window and adjust the image so that it is centered on the pelvic region.*

➤ *Adjust the Layer Indicator to 3, and click the Highlight button in the Tool palette.*

➤ *In this image the skin has been removed, revealing the female external genitalia.*

1. Use the Identify tool to label the structures indicated in the following diagram.

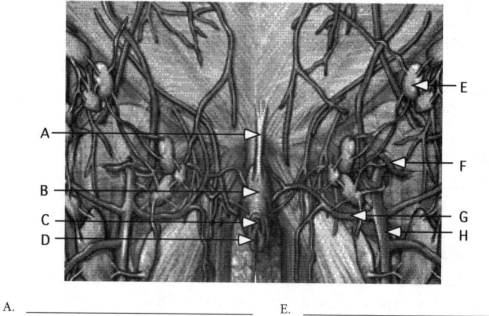

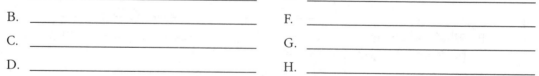

A. _____   E. _____
B. _____   F. _____
C. _____   G. _____
D. _____   H. _____

Exercise 11.6 3D Female External Genitalia

➤ *Close the Dissectible Anatomy window and select Open Content from the File menu. Click 3D Anatomy and select Female Reproductive. Click Open.*

➤ *Maximize the view on your screen.*

1. Use the Structure List for 3D Female Reproductive at the top of your screen, scroll down the list, and identify the following structures on the diagram below. (Note: The structures are listed in alphabetical order.)

- Clitoris—Skin
- External orifice of urethra—Skin
- Hymen—Skin
- Labium majus—Skin

- Labium minus—Skin
- Prepuce of clitoris—Skin
- Vestibule of vagina—Skin
- Vulva—Skin

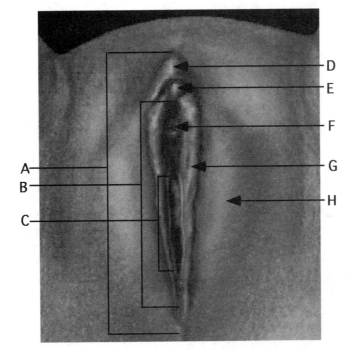

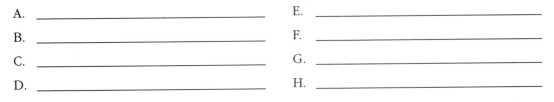

A. _____   E. _____

B. _____   F. _____

C. _____   G. _____

D. _____   H. _____

Exercise 11.7 Internal Genitalia

➤ *Close the 3D window and open Dissectible Anatomy. Select Female and Anterior and center the image over the pelvis.*

➤ *Adjust the Layer Indicator to 206. The uterus is anterior to the rectum.*

1. What covers the uterus in this view?

   _____

2. Identify the large vertical vein shown on the right side of the screen just lateral to the superior ramus of pubis.

   _____

3. You will see a smaller forked artery and vein appearing inferior to the ovary.

   a. Identify the superior branch: _____

   b. Identify the inferior branch: _____

Beyond A I A

4. What is the function of the peritoneum?

   _____

➤ *Set the Layer Indicator to 208. In this image, the distal end of the large intestine has been cut.*

5. What is the name of the pink organ lying anterior to the rectum within a plexus of arteries and veins?

   _____

6. What is the name of the network of veins around the uterus?

   _____

7. Lying just superolaterally to the fundus of the uterus and rectum is a similar network of yellow fibers. What is the name of this network of yellow fibers?

   _____

8. On the far sides of the image, two large vertically oriented blood vessels appear, a medial vein and a lateral artery.

   a. Identify the medial vein: _____

   b. Identify the lateral artery: _____

9. Set the Layer Indicator to 223. What is the name of the long, slender artery that runs behind the infundibulum onto the front of the ovary?

   _____

Beyond A I A

10. What is a plexus?

    _____

11. What is the direct blood source of the ovarian arteries?

    _____

12. Set the Layer Indicator to 226. Use the Identify tool to label the following diagram.

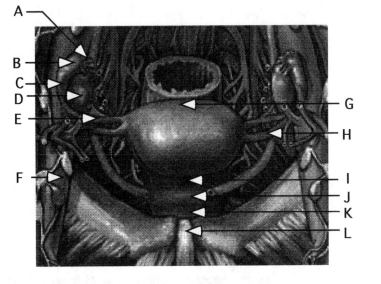

A. _____	G. _____
B. _____	H. _____
C. _____	I. _____
D. _____	J. _____
E. _____	K. _____
F. _____	L. _____

Beyond A I A 13. What is the function the fimbriae of the uterine tube?

_____

14. What is the function of ovaries?

_____

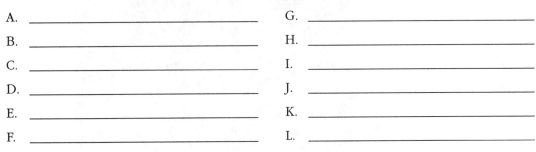

(Note: Tubal ligation. Birth control in females can be achieved surgically by excising a small segment of the uterine tubes at the isthmi and tying off or cauterizing the remaining cut ends. This prevents ova from entering the uterus or meeting with spermatozoa. Two small incisions are made in the abdomen, and the procedure, which is known as *tubal ligation*, is performed with the aid of a laparoscopic device.)

(Note: Cervical cancer. Cancer of the cervix is the third largest killer of women in the United States. This cancer starts as cervical dysplasia, an abnormal change in the shape and number of cervical cells, and then progresses to malignancy. Some evidence links the incidence of this cancer to infection with the male papillomavirus that causes genital warts. Early diagnosis of cervical cancer can be made with a Pap smear. In this test, cells from the surface of the cervix and vaginal wall are scraped away and examined microscopically for abnormalities. The test is named for Dr. George Papanicolaou who developed the procedure. Yearly Pap smears are now recommended for all women 18 years of age and older [younger if they are sexually active], along with a test for the papillomavirus. When three consecutive Pap smears show normal results, the interval between tests is increased to 2 or 3 years.)

Clinical Animation

To observe a clinical animation of egg cell production, do the following:

➤ *Select Open Content from the File Menu. Select Clinical Animations. Select Reproductive from the Body System drop-down menu. Select Egg Cell Production. Click the Open button. Expand the window by clicking the Maximize button.*

➤ *After the animation finishes, close the window.*

Exercise 11.8 Female Reproductive Anatomy (Medial View)

➤ *Select Medial from the View button drop-down menu.*

1. Set the Layer Indicator to 47. Adjust the image to match the diagram below, and label the indicated structures.

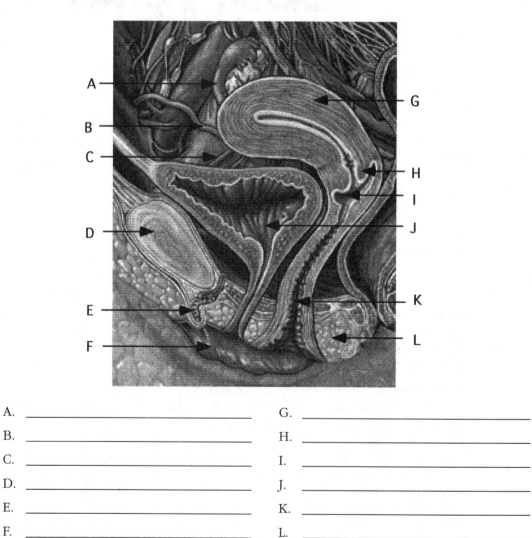

A. _____   G. _____

B. _____   H. _____

C. _____   I. _____

D. _____   J. _____

E. _____   K. _____

F. _____   L. _____

**Beyond
A I A**

2. In this midsagittal section, the interior cavity of the uterus is lined with a white layer of tissue. What is this interior uterine lining called?

_____

(Note: Endometriosis. Sometimes, uterine lining cells can escape from the uterus to the pelvic cavity and reproduce, causing a condition known as *endometriosis*. The cells are still under the same monthly hormonal control of menstruation and proliferate in areas where they can cause pain, inflammation, and, if growing inside the uterine tubes or on the ovaries, infertility.)

Exercise 11.9 Mammary Gland (Anterior View)

➢ *Select Anterior from the View button drop-down menu.*

➢ *Adjust the image until the chest is centered. Set the Layer Indicator to 6. Click the Normal button to return to full color.*

1. What is the name of the grayish-white structures that radiate outward from the nipple?

_____

2. What is the name of the light green structures that are found over the breast?

_____

3. What is the name of the network of lighter gray or white lines that covers the breast (farther away from the nipple) and consists of fibrous strands that attach the breast to the overlying skin?

_____

4. The breast is shown to be made up primarily of pink cell-like bodies. What are they called?

_____

**Beyond
A I A**

5. What is the function of lactiferous ducts?

_____

6. What is the function of mammary glands?

_____

➢ *Select Lateral from the View button drop-down menu.*

➢ *Set the Layer Indicator to 2.*

➢ *Without closing the current figure, select Open Content from the File menu. Choose Dissectible Anatomy, Female and Anterior. Click Open.*

➢ *Enlarge the window and center on the breast area. Set the Layer Indicator to 6.*

➢ *Select Tile Vertically from the Window menu.*

7. Adjust the images in both windows so that they match the diagram that follows. Use the Identify tool to label the diagram.

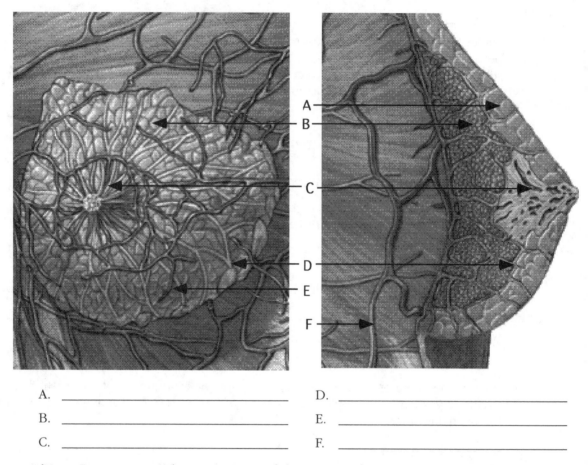

A. _____   D. _____

B. _____   E. _____

C. _____   F. _____

(Note: Breast cancer. The connections of the intercostal veins with the vertebral venous plexuses provide a route for the spread of cancer cells [metastasis] from the breast to the vertebrae and then to the skull and brain. When carcinoma [CA] of the breast spreads to the retromammary space [region of loose connective tissue between the breast and underlying pectoralis major muscle], contraction of the pectoralis major muscle causes the breast to move superiorly, a clinical sign of advanced malignancy of the breast. The lymphatic and venous drainage of the breasts is important in the spread of breast cancer. Cancer cells are carried by the lymphatic vessels of the breast to lymph nodes, particularly to the axillary nodes. Cancer cells will lodge in these nodes, producing groups of tumor cells called metastases. Enlargement of these nodes in a woman suggests the possibility of breast cancer.)

Exercise 11.10 Female Pelvis

➤ *Close the "Female Anterior" and "Female Lateral" windows.*

➤ *Select Open Content from the File menu. Choose Atlas Anatomy.*

➤ *Select Pelvis and Perineum from the Region menu, Reproductive from the System menu, and Inferior from the View menu. Select the "Female Pelvic Diaphragm (Inf)" thumbnail icon. Click Open.*

➤ *Expand the window.*

1. Label the lettered pins on the following diagram.

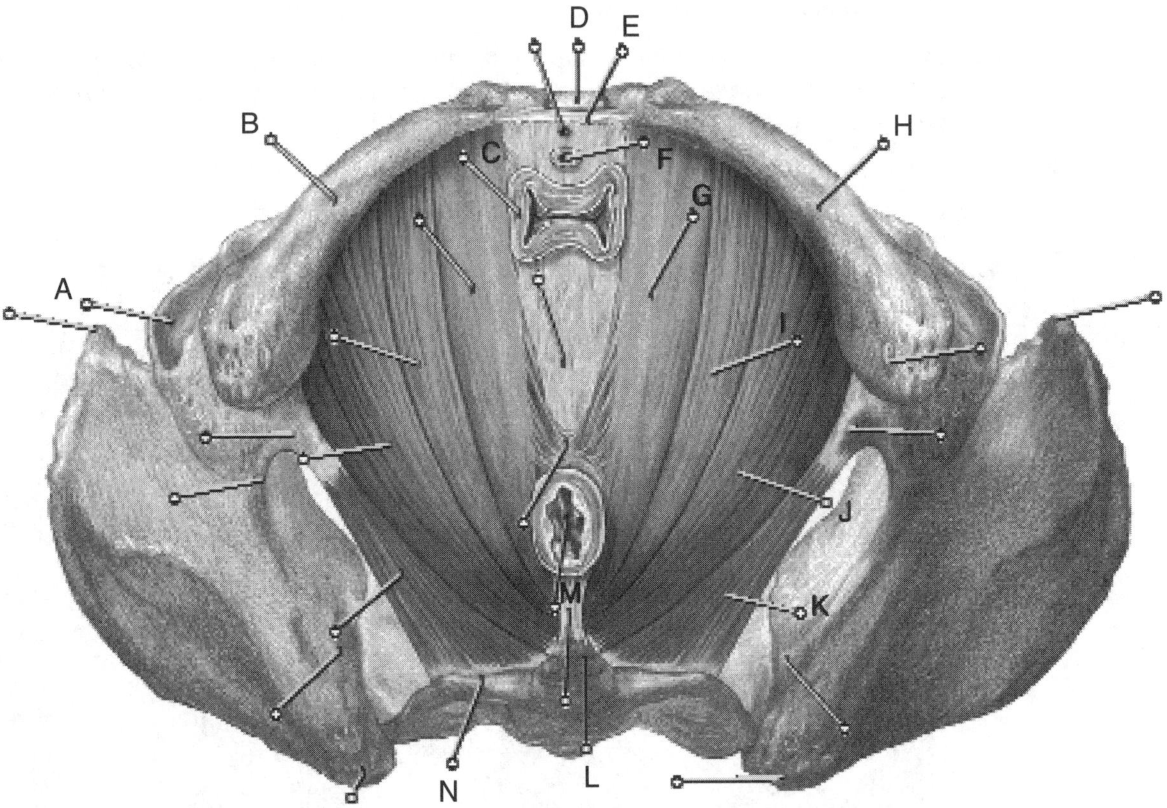

A. _____   H. _____

B. _____   I. _____

C. _____   J. _____

D. _____   K. _____

E. _____   L. _____

F. _____   M. _____

G. _____   N. _____

Beyond A I A

2. With what bony structure on the femur does the acetabulum articulate?

_____

3. What is the literal meaning of *levator ani*?

_____

4. The levator ani forms a sling that supports the pelvic viscera. What is the name of this muscular sling?

_____

5. In females, what is the name of the region between the vagina and anal canal?

_____

6. From what organ does the urethra arise?

_____

➤ *Close the "Female Pelvic Diaphragm (Inf)" window.*

➤ *Select Open Content from the File menu. Choose Atlas Anatomy.*

➤ *Select Pelvis and Perineum from the Region menu, Reproductive from the System menu, and Medial from the View menu. Select the "Fascia in Female Pelvis (Med)" thumbnail icon. Click Open.*

➤ *Expand the window.*

7. Label the diagram that follows.

A. _____   F. _____

B. _____   G. _____

C. _____   H. _____

D. _____   I. _____

E. _____   J. _____

8. List the names of the three layers of the uterus from inside to outside.

a. Inner: _____

b. Middle: _____

c. Outer: _____

9. The male penis is said to be homologous to which female structure?

_____

10. What is the literal meaning of the term *myometrium*?

_____

11. What is the significance of the myometrium?

_____

12. What is the common name for the uterus?

_____

13. Which structure represents the longest portion of the birth canal?

_____

➤ Close the "Fascia in Female Pelvis (Med)" window.
➤ Select Open Content from the File menu. Choose Atlas Anatomy.
➤ Select Pelvis and Perineum from the Region menu, Reproductive from the System menu, and Anterior from the View menu. Select the "Female Pelvic Organs (Ant)" thumbnail icon. Click Open.
➤ Expand the window.

14. Label the lettered pins in the following diagram.

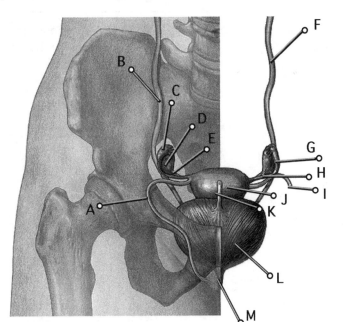

A. _____   H. _____

B. _____   I. _____

C. _____   J. _____

D. _____   K. _____

E. _____   L. _____

F. _____   M. _____

G. _____

15. The round ligaments are narrow, flat bands of fibrous connective tissue. To what structure do they attach at their:

    a. superior end? _____

    b. inferior end? _____

16. What structure anchors the ovaries to the uterus?

    _____

Beyond
A I A

17. What is the function of the round ligaments?

    _____

    _____

18. What is the function of the infundibulum?

    _____

    _____

19. What structure represents the widest and longest portion of the uterine tube?

    _____

20. What is the function of the urinary bladder?

    _____

    _____

21. What particular type of epithelial tissue lines the inner surface of the urinary bladder and urethra, and what is the function of this type of epithelium?

    _____